The Cultural Dynamics in Water Management from Ancient History to the Present Age

The Cultural Dynamics in Water Management from Ancient History to the Present Age

Edited by

Xiao Yun Zheng

IWA PUBLISHING

Published by **IWA Publishing**
Unit 104–105, Export Building
1 Clove Crescent
London E14 2BA, UK
Telephone: +44 (0)20 7654 5500
Fax: +44 (0)20 7654 5555
Email: publications@iwap.co.uk
Web: www.iwapublishing.com

First published 2021

British Library Cataloguing in Publication Data
A CIP catalogue record for this book is available from the British Library

ISBN: 9781789062038 (paperback)
ISBN: 9781789062045 (eBook)
ISBN: 9781789062052 (ePUB)

Contents

About the Editor

Dr. Xiao Yun. Zheng is a water historian and anthropologist, the distinguished professor at School of History and Culture, Hubei University of China. Currently, he is also holding the offices of president of China Institute of Yangtze River Culture Studies and director of Institute of Water History and Water Civilization in university. The areas of his research and education include river history, urban water history, water culture, water anthropology, hydro-engineering in ancient civilization, etc. His study based on China but also include a global comparative prospect. He has been conducted academic activities and fieldworks in more than thirty countries. He is past president of International Water History Association, past vice president of Chinese Senior Expert Committee for Water Culture, China Ministry of Water Resource. Currently, he is the vice president of China Water Museum Alliance, Scientific advisor of Water Museum Global Network, official advisor of Water Culture Institute, USA, etc. and visiting professor in several universities in China and international. He has published 13 books and 150 research papers in China and abroad. He was awarded the title of Honorary Doctor by Thessaloniki Aristotle University in 2017.

List of Contributors

Andreas N. Angelakis National Agricultural Research Foundation, Institute of Iraklion, Iraklion, Greece.

Seyed Ali Badri Department of Human Geography, Faculty of Geography, University of Tehran, Tehran, Iran

Nguyễn Chí Bền Vietnam National Institute of Culture and Arts Studies, Hanoi, Vietnam.

Eriberto Eulisse Civiltà dell'Acqua International Centre, and Ca' Foscari University of Venice, Italy.

David Groenfeldt Water-Culture Institute, Santa Fe, USA.

Nguyen Thi Hien Vietnam National Institute of Culture and Arts Studies Hanoi, Vietnam.

Petri S. Juuti History of Technology (Oulu) and Finnish History (Turku), Build Environment, Tampere University, Finland.

Majid Labbaf Khaneiki Geography and Rural Planning, Department of Human Geography, Faculty of Geography, University of Tehran, Iran

Du Longjiang China Institute of Water Resources and Hydropower Research, Beijing, China.

José Ricardo Martínez Museography, Yaku, Water Museum, Ecuador.

Jodie Matthews Canal & River Trust and University of Huddersfield, UK

Riikka Rajala History of Technology (Oulu) and Finnish History (Turku), Build Environment, Tampere University, Finland.

Cristina Reyes Yaku, Water Museum Park, Ecuador

Mohammad Reza Rezvani Department of Human Geography, Faculty of Geography, University of Tehran, Tehran, Iran

Ansari Taha National Agency of Hydraulics Resources, Adrar, Algeria.

Mari Takeda Department of Agricultural and Resource Economics, The University of Tokyo, Japan.

Vu Hong Thuat Vietnam Ethnology Museum, Hanoi, Vietnam.

Emilia Vallejo Yaku, Water Museum Park, Ecuador

Section 1

The Theoretical Discussion

Chapter 1

Water culture: The theoretic structure and dynamic mechanism

Xiao Yun Zheng

ABSTRACT

Since water is comprehensively connected with human society, diversified way of thinking, using, and managing water have developed. This has led to the diversification of water culture worldwide. Water culture was generated from water management, the social demand, and the spiritual feel of water. Consequently, it has also influenced water management comprehensively beginning from the earliest times to the present age. People worldwide rely on their own ideas, values, social relationships, institutions, and religion to manage water. This, however, has made water culture an important dynamic in water management and has also made water management acceptable, effective, equal, and sustainable. Water culture also deeply influences the spiritual world and social life of mankind. Water culture has a structural make up involving the levels of ideas, institutions, social life, and physical factors and is based on the attributes of nationality, civilization, local, and times. This theoretical structure and attributes provide a better understanding of water culture based on scientific and holistic dimension. Meanwhile, a dynamics mechanism was modeled as Double Dynamics and Bidirectional Dynamics to explain the relationship between water and mankind so as to better understand how the cultural generation is driven by water and the cultural initiative in water management. This chapter aims to build the structure and define the types of water culture, to exploring the dynamic mechanism of water culture in water management and human society, and to understand the generation and formation of water culture.

Editor: Xiao Yun Zheng
doi: 10.2166/9781789062045_0003

Keywords: water culture, theoretical structure, cultural generation, cultural attribute, dynamic mechanism, water management

1.1 INTRODUCTION

Water is a very important necessity for life. This means that water management has been a foundation for human life and development throughout human history. Human beings have different demands and purposes in managing water, such as daily life, sanitation, agriculture, industry, transportation, urbanization, and military, political, social, religious, and artistic purposes. However, among these demands and purposes, water related culture or water culture is a more foundational factor that is interrelated to any objective of water management. Water management has driven the generation of culture, while culture has influenced water management. People from different regions, ethnic groups and civilized backgrounds depend on their ideas, values, customs, social relationships, norm, and institutions and religion to manage water. The cultural factor also determined how water was managed. It has been broadly demonstrated that huge human cultures were generated that were related to water or driven by water. In addition, water culture also featured in human diversity due because it was shaped in different social, ethnic, and religious backgrounds and geographic conditions globally. Hence, it is important to consider the cultural influences on water management from a local area to a region, from individual to collective, and from an ethnic group to all human beings.

Water culture also has some features that are related to environmental and resource management, and to sustainable development. When people touch water, use water or manage water, water related culture is generated. After the culture has been generated, it becomes a foundation of water management. On the other hand, people also manage water depending on their values, religious idea, social intention, and institutions to establish relations between water and the societies. Water culture is also generated through this process. In addition, since water culture was generated in different environments and societies, it is characterized by diverse natural and social conditions, although at the same time, it displays many common features in a region. This is due to common demands and purposes since water is an indispensable substance for human life. More so, some cultural factors concerning water go beyond the ethnic and local societies; for example, religious factors may also influence the features of water culture.

Culture is a multifaceted aspect of human society. This is seen in spiritual phenomena, customs, behavior, lifestyle, and physical constructions. Hence, in order to understand the cultural influence in water management, modeling the water culture is important. Modeling and structuring water culture will provide a better understanding and survey of its dynamic role at specific levels. Furthermore, the dynamic role for water management can be adjusted depending on demands. It should be structured from the levels of ideas, behavior, and institutions to physical outcomes. It is also important to understand how water culture has been influencing or determining water management depending on prevailing situations (Xiao Yun, 2013a). Nonetheless, water culture has played an important dynamic role in water management from history to this present age, and it will continually be a key factor for water future. Therefore, there is a wide scope for understanding the dynamic role of water culture in history, exploring the way to apply water culture, and increase the effectiveness of cultural dynamics in water management.

Currently, human society is experiencing common environmental emergencies, such as coping with the current global water crisis and maintaining water sustainability. Therefore, it is necessary to maintain and rebuild the common culture for water use and sustainability. In addition, understanding the complexity and mechanism of the source of water culture will provide a better understanding of the dynamic role of water culture. At the same time, it is essential to positively drive the generation of water culture for water sustainability.

1.2 THE STRUCTURE AND ATTRIBUTES OF WATER CULTURE

What is the definition of water culture? In the most basic understanding, water culture is the water related culture which was mainly generated from interactions between water and human beings, and it was also formed by 'other' cultural factors such as the religious meanings to water that shaped the secret meaning of water. Since the definition of water culture is still limited in discussion, water culture can be defined as the cultural complex of water related culture which has spiritual, social, institution, and material outcomes. On account of this, culture does not just exist solely in the human spiritual world or as one aspect of society, but it is multifaceted in its existence in human society. Subsequently, water culture could be considered as a cultural system which is made up of structured levels and elements. At the same time, water culture is generated and clothed in diversified social and natural environments. Thus, water culture is also diversified in human society and influences water management conditionally.

For a clear understanding of water culture scientifically and holistically, it is important to establish the structure and discuss its social and natural attributes.

1.2.1 The structure of water culture

Since water culture is not just considered as a cultural phenomenon but as a system, it is essential to theoretically structure it into four levels as shown below:

1.2.1.1 The level of ideas

This level comprises the worldview, value, knowledge, and aesthetics of water. The worldview includes the understanding of the role of water in shaping the universe, the meaning of water, as well as the belief, worship, inspiration of water, etc. The value consists of the judgment of relations of water, human survival benefits, and identity of water. Knowledge includes the understanding of the nature and culture of water and the management of water in daily life and in other activities. The aesthetics includes hobbies, feelings, and sentiments about water as well as other spiritual elements of water, such as the religious understanding of water.

1.2.1.2 The level of institution

This level is made up of the norms, rules, and regulations for use, management, and control of water. Based on the conditions, the institutions formed for water management carry out water management in an orderly way which people follow.

1.2.1.3 The level of social life

This level encompasses the following: social behavior, customs, social relationships, lifestyles, etc. The social behavior includes behavior regarding daily water use, which depends on habits and the collective behavior pattern for water. Customs consist of water related customs, such as personal rituals related to water, water use in festivals, water rituals and water storage, etc. The lifestyle takes account of the facets of life are based on water, such as daily water use for cooking, washing, bathing and sanitation, and working with water. Social relations comprise the social associations formed by water matters. As a result, people could be organized and this has formed social life, responsibilities, and benefits.

1.2.1.4 The level of physical constructions

This level includes the physical constructions of water management, whereby the hydro-constructions connect with cultural meanings and symbols in the process of use, water treatment, and beautification of the environment, such as canals, channels, wells, dams, waterscapes, hydro-facilities (Xiao Yun, 2013b).

According to the description given above, water culture has a structure from the levels of spiritual to social, institutional, physical as well as a series of elements which makes up the structure. It is systematic but the composed elements could be surveyed. Accordingly, water culture is not an elusory phenomenon. According to this constitution, water culture is a comprehensive system that would be considered as a subculture of human culture. Confidently, this suggestion will help to determine the value of water culture in the human cultural prospect.

1.2.2 The attributes of water culture

Water culture has been generated and housed depending on social, cultural, and environmental condition. It has general features in wider human society, but it is also characterized by social cultural conditions and local environments. When thinking about the cultural dynamic influence of water management, it is important to ponder upon it at a specific level involving the element as well as specific attributes, such as the geographic, social, cultural, or time attributes. This will help to provide a better understanding about water culture in a holistic view. Generally, water culture could be classified into four attributes:

1.2.2.1 The attribute of nationality

National societies are a reality in human society. Water culture is also housed in multiple national societies, which are characterized by social and cultural features, nationally and ethnically. Accordingly, people always depended on their national or ethnic culture to understand and manage water. This means that they depended on their worldview, religion, social values, and norms to manage water. Thus, this brought about a different understanding of the use of water between the societies and formed the cultural diversity of water. It is the most vivid feature of water culture.

1.2.2.2 The attribute of the civilization

Water culture was generated in different societies. At the same time, it was also influenced by or characterized by the civilizational backgrounds, regionally, nationally, and religiously. For example, Chinese water culture was deeply influenced by Taoism. On the other hand, Japanese water culture was influenced by Shintoism. Furthermore, in many other nations, water culture is influenced by Buddhism, Christianity, Islam, etc. Furthermore, it is also characterized based on the national or regional civilizations such as the socially organized characteristics, the hobbies, aesthetics, even the sentiments related to the water, etc.

1.2.2.3 The attribute of the local

Since humankind lives in diversified environments, varied ways to use water and live with water were adopted according to the local environmental conditions, such as the coast, the river, the lake, the desert. This has shaped the difference of the water culture locally.

1.2.2.4 The attribute of time

Water culture was generated over different historical time periods. Therefore, it has been characterized by the social, cultural, and other conditions of those times. Thus, water culture may be considered successional, but it is also characterized with features of past times.

1.2.3 The types of water culture

The types of water culture are also important in order to understand the features of water culture. Water culture could be classified into three types depending on its source:

(1) Water culture generation is driven by adapting to the local water situation and water management. During this process, the ideas, values, customs, lifestyles, beliefs, behavioral forms, norms, and social relations which are related to water should be generated. The lifestyle related to water locally should be shaped in accordance with the water situation, for example, living near a lake or a river, or in an arid place.

(2) Water culture is shaped by the cultural intentions. Water is taken as a substance to express cultural intentions, and then it is shaped to relate to water cultures. For example, water was given secret meanings by the religions and used in rituals to express religious intentions. This is a popular phenomenon worldwide. Water is also usually endowed with other cultural meanings and uses a cultural matter, for example, water is considered as a supernatural force to use for medical treatment.

(3) Water culture is also generated from intentions-based water management. Basically, water culture depends on values and intentions to be managed. For example, the social norms, rules, customs, and behaviors for water management were directed by values, ethics, identity, etc. Currently, sustainable environment values should direct societies to shape friendly social behavior for water use, saving water, etc.

In addition, since water culture was generated in different regions, societies, and civilizations, it is also reflected in diversity. It is characterized in different societies against the background of history, tradition, ethnicity, religion and civilization, etc. This situation brings difference of water culture such as in the forms, structures, categories, and scale in human societies. Basically, water cultural diversity as a traditional form was generated from the historical process of society and has played a significant role in water management historically and locally.

The theoretical structure and attributes of water culture provides a better understanding of water culture in scientific and holistic dimensions. Water culture should be considered as a multifaceted and multielement system, and it is clothed in different attributes. Water culture is not just an abstract phenomenon but also has concrete form in human societies. In addition, the levels and the elements of water culture interact socially, environmentally, and historically. Accordingly, water culture actions in water management are based on the proviso that not all elements of water culture can influence water management. Thus, it needs to be analyzed specifically to ascertain which elements play a significant role in water management, for example, religion plays a significant role in water management in some societies but the role of social norms might be emphasized in others.

Also, water culture is a system which dynamically and changeably depends on conditions and times.

1.3 THE CULTURAL DYNAMICS MECHANISM IN WATER MANAGEMENT

Since the structure of water culture has been modeled, where water culture comes from and how water culture plays a dynamic role in water management are key questions that need to be answered. It is important to model the cultural dynamics in water management to explain how cultural generation is driven by water and to better understand cultural initiative in water management. This will provide a clearer view about the cultural mechanism in water management, and this mechanism can be rebuilt for water sustainability.

The dynamic model of the human–water relationship has been built as the Double and Bidirectional Dynamics Structure. This helps in explaining the interactional mechanism of cultural dynamics and water management. Consequently, the details of the model are described as follows.

1.3.1 The double dynamics structure

This means that there are two aspects of the dynamics of water management. The first is the demands dynamic. Human beings depend on their demands of life, substance, residence and economic development to manage water. This includes the management of water for urban water supply, agricultural irrigation, water transportation, and water use for daily life.

The second aspect is the cultural dynamic. Human beings depend on their demands to manage water for cultural purposes, such as ideas, values, aesthetics, religions, political intentions, customs, social norms, and hobbies.

1.3.2 The bidirectional dynamics

As stated above, the demands are driven by water management leading to creation of the culture. At the same time, people also depended on culture to manage water, and this brings cultural factors to water management. Hence, culture influences water management and becomes a dynamic of water management. Thus, water management and culture interact in a bidirectional process.

People popularly depended on their culture and civilized backgrounds to manage water worldwide. Therefore, culture is considered as a foundation of water management. Culture is not a passive factor in water management but would determine water management on many occasions, for example, the ideas of water may determine how to use water and the social norm may determine the form of irrigation chosen. Therefore, cultural purpose-driven water management is an independent dynamic factor. However, the extent of cultural influence and determination on water management depends on complex social conditions:

culture influences water management to different degrees. In addition, the dynamics of demands and culture also interacts and influences each other. The demands of dynamics have driven the generation of water culture, while culture has also driven the generation of new demands for water management.

Originally, parts of culture-related water were generated from demands of water management. The practices of water management brings the generation of water culture, for example, the idea and value of water, the knowledge system of water, social norms for water use, the water related arts, etc. However, it is worthwhile noting that many cultural factors related to managing water were not generated by water management but influenced such management. This includes some religious factors, social and politic ideas, values, etc. People may depend on their social values to distribute water or on religious ideas to protect water resource. These cultural factors, which usually originate from the social value system in the society, are not derived from water practices but are influenced by water management in varying degrees.

Culture is not just influenced by water management but has also created the new demand of water, for example, the sanitation facilities and the artificial waterscapes construction which depended on the cultural ideas are the popular factors promoting water management in many societies. In this case, a circular mechanism which was formed between the dynamics of demand and culture would be repeated as: the demands and culture driven by water management, water management driven by the generation of culture, and the culture driven by water management and generation of water demands.

Consequently, it is important to understand the cultural dynamics of water management. This also could be based on the mechanism to use cultural means to improve water management in this present age since it is a fact that technical means has currently made significant progress in water management. At the same time, culture is insubstantial and results in ineffective water management making it difficult to cope with the water crisis in this present age. Furthermore, good water management primarily depends on human cultures, which include good ideas, values, understanding, behavior and social institutions related to water. A lot can be learnt from history in the global communities and also from current situation to rebuild a good culture for sustainable water management.

1.3.3 How culture is generated by water management

Water culture was mainly generated from water management, but it is a complex process at the different levels of the society.

Firstly, it is necessary to discuss how water culture is generated from water management. Structured below, water culture is divided into four levels: the ideas, institutions, social life, and physicals. Thus, the cultural generation discussed here will follow this structure:

1.3.3.1 The level of idea

The level of idea includes the knowledge related to water and the understanding, values, beliefs, and aesthetics of water. It can be reflected by worship, religion, myth, arts, social norms, etc. Accordingly, it is a vivid and influential level of culture. Ideas could direct and influence social behavior, institutions, and norms for water management. However, the idea of water came mainly from living experiences and practices of water management, which was usually a long process historically. For example, the case of Da Yu is typical in China. According to legend, Chinese state history began when Da Yu controlled the floods. It was said that at that time (about 2000 BC), the land was often flooded and people suffered many disasters due to the uncontrolled floods. Da Yu, a son of the state leader, was appointed to lead the people to manage the floods by channeling water into the rivers and finally into the east sea, whereas his father had managed floods by using dams to stop water. When flood water flowed into the sea through rivers and was controlled effectively, the people were saved from disaster. Accordingly, the state grew quickly to be the first Chinese dynasty, that is the Xia Dynasty. Da Yu was recommended as the emperor of the Xia state. This story and its implications also deeply influenced Chinese society and water management. An important principle learnt from this event is that water needs to be controlled but not stopped, and to manage a society, it would be better to follow this principle. Thus, the best water management practice is to channel water to flow regularly. On the other hand, the legend also created a comprehensive cultural phenomenon in China. Da Yu was also acknowledged as the god of water management. The Da Yu temples were built and were places of worship in China, Japan, and Korea (Wang, 2017), and some Southeast Asia countries. It is difficult to demonstrate that this legend was a true event. However, this does not dispute the fact that it reflects the realistic interactions of water and the people in ancient time. It has also generated rich elements of water ideas which the Chinese people have kept in mind. Therefore, this story reflects the point that ideas about water sprang up from realistic water practices.

Another topical case in China is the Dragon King belief. This dates back to about 1000 BC according to the historical documents. According to the legend, the dragon sometimes lives on land or in the springs, rivers, and lakes and controls water resources. When it lives in the sky, it always hides in the clouds to control rain. Therefore, the rainfall from the sky and the water on land are controlled by the dragon. The dragon was not just a god of water but also the most powerful animal in the Chinese spiritual realm, and this was the symbol of the royal family for 2000 years, from the Qin dynasty that first unified China to the final dynasty Qing. Dragon King temples were built close to rivers, lakes, and springs. For example, it is a very popular belief that the spring is the home of the dragon. Thus, it was very common to build a Dragon King temple close to a spring, and then the spring was often called 'dragon pool' (Figures 1.1 and 1.2). This led to

Figure 1.1 The Nine dragon pool in Yuxi city, China.

Dragon worship becoming a comprehensive cultural phenomenon in ancient China. The dragon culture not only deeply influenced water management, but it also influenced the social and political life in ancient China. The belief of the dragon reflected the relationship between water and people, as well as the importance of water management. In ancient times, people in central China lived in water-rich circumstances to control floods, improve waterscapes for habitation, develop irrigation for agriculture, etc. The history of life then was a history of water management. The dragon becomes the symbol of power because the dragon is the legendary animal that controls water, which is the essential resource of human

Figure 1.2 The statues of Dragon King and the queen (with permission of Xaio Yun Zheng).

life. Controlling water brings power, and this was an essential conception in ancient Chinese society.

Water management has also driven the generation of values, ethics, worldviews, and knowledge related to water. In southwest China and the Mekong River basin, there are many ethnic people who believe that the world was formed from water and the real world was also generated from water (according to the myths of the Tai-Dai, Hani ethnic peoples). There are also myths about how the god of water controlled water and influenced human beings. Based on the natural environment in this region, myths, values, and the knowledge system of the local people were rooted in the most common subsistence farming, which is rice cultivation, again reflecting the understanding of water. At some point, the myth is not just seen as a story but as an influence in people's idea and value for water management (Xiao Yun, 2013c). For example, the Tai-Dai ethnic group is a nation distributed in Yunnan province of China and in Myanmar, Vietnam, Laos, and Thailand. The Tai-Dai society focused on wet rice cultivation, and water is the foundational material for their subsistence. Accordingly, the Tai-Dai people strongly respect water and water played an important role in shaping their spiritual world, knowledge system, values, and social life. Depending on the demands of water management for irrigation for rice paddy, many social norms and customs were formed. At the same time, people believe that water is not just the essential element for life and subsistence, but it is also the material that could purify the world and the soul. Therefore, water is widely used in religious and social rituals to express blessing, avert disaster, purify the soul, contact the deity, and so on. Many social and personal ethics and norms were shaped depending on their long practice of water management.

1.3.3.2 The level of social life

People live within the local water environment and touch water every day. Thus, social life was unavoidably influenced by water. On the one hand, people have to adapt to the local water environment, allowing water related lifestyles and custom to be shaped. On the other hand, people may change the natural water situation to meet their demands for better residential condition and development. This is followed by a new social situation coordinated with water. Culture could be generated in this process, and this process has been broadly demonstrated.

Tuo Shan Yan Dam is a small dam which was built above Yinjiang River at Yinjiang Town, Ningbo City, Zhejiang of China in 833 AD. It has greatly contributed to Ningbo City, which is currently the biggest port city of China. It is located at the midpoint of the Chinese coastline and the south wing of Yangtze River delta. The dam and the corresponding systems were the key hydro-infrastructure for the city's existence and development. In addition, there were irrigation activities in the countryside near the city from the 9th century up to 1987. A large dam was built on the upper reach to control the water of the river. The dam played an important role in local water management and met a series of objectives such as regulation of river water, preventing floods, stopping the salinization, urban water supply, storage of freshwater, irrigation, and ship transportation. A remarkable phenomenon is that a popular culture related to the dam was generated after its construction. This deeply influenced the local people's spiritual world and social life. After the dam had been built, the local people were saved from a salinization disaster and agricultural productivity was effectively increased as irrigation also improved. Furthermore, some towns were built along the rivers as well. The most important improvement is the supply of water to Ningbo City, and this brought about stable circumstance permitting the city's long development. All of this was embedded in the local spiritual world and practiced in real life by means of memory, gratitude, and worshipping the builders of the dam for a thousand years. According to the historical record, the dam was built and organized by the governor of the local county, Wang Yuan Wei in 831, and ten young male workers died during the construction. The local people celebrated the construction and every year they showed their gratitude to governor Wang and the ten young men who paid with their lives during the construction. Governor Wang and the ten young workers were recognized as gods of the water. According to the Chinese calendar, every year, the local people celebrated three times at the dam and the celebration became a temple fair. Thousands of people from the local and neighboring areas gathered together at Yin Jiang town for the temple fair: their activities included worship, selling foods and handicrafts, and dance, parades, etc. This phenomenon also influenced a wide area in the east Zhejiang province. There are more than 30 temples named Tuo Shan in east Zhejiang province dedicated to worship of Governor Wang (Tua, 2011). The worship of Governor Wang and the water gods keeps alive the memory of people's gratitude to the dam constructors. This is also a way to educate the people to understand the water history of the local for better water management. Furthermore, the worship of the water gods, Wang, and the ten people who died during the construction has formed many moral principles to educate people to have better personal culture, for example, good service to the public, to be cultured people or to be a good officer who cares for the common people like Wang. Culture was generated from the dam construction and local water management and has influenced the water management until recently. This has formed the knowledge and norms, institutions for water management, and also broadly influenced local social life (Xiao Yun & Min, 2017) (Figures 1.3 and 1.4).

The transformation of a fountain is another typical case to demonstrate how culture was generated from water management. Generally, a fountain is a piece of architecture which pours water into a basin or jets it into the air to supply drinking water and/or for a decorative or dramatic effect. Fountains are known since the early Bronze Age in southeastern Greece and in Indus valley. Thus, it was the terminal point of water supply from springs or other sources by pipes, or

Figure 1.3 (a) Tuo Shan Yan Dam. (b) Tuo Shan Yan Dam, with permission of Xaio Yun. Zheng.

aqueducts, to provide water for drinking, bathing, and washing for the residents of palaces, cities, towns, and villages. The advance hydro-technologies including fountains, which developed in Crete, the center of Europe's first advanced civilization, the Minoans, was mainly used for providing drinking water (Angelakis, 2015). However, it was further improved and expanded to other European and Mediterranean regions by Mycenaeans, Etruscans, people of Classical and Hellenistic Greece, Romans, Venetians, and Ottomans (Petri *et al.*, 2013). In the present day, fountains are found worldwide and continue to serve as sources of drinking water in some places, for example in many rural places. However, the form and function of the fountain had been widely transformed in ancient times. A fountain is not just a source of drinking water but also has individualized and diversified meanings, for example, power, religion, monument, symbol, art, decoration, enjoyment, etc. During the Roman era within the first century, the fountain was at a glorious age. Fountains not only served as

Figure 1.4 (a) Parade with Wang's statue. (b) Parade with the ten men statues (Ningbo News, 2014).

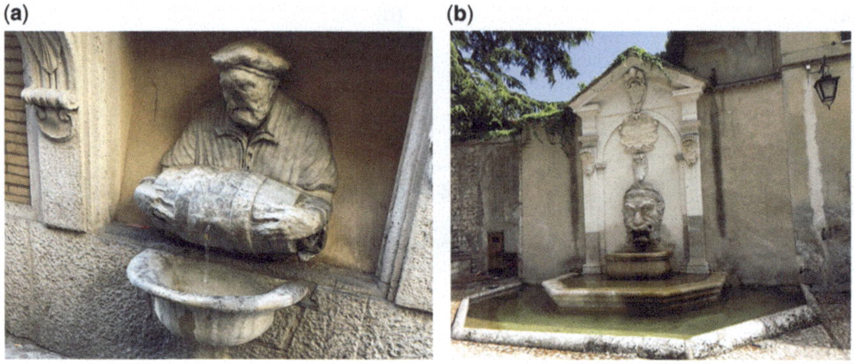

Figure 1.5 (a) An ancient fountain for drinking water in Rome City. (b) An ancient decorative fountain in Rome City, with permission of Xaio Yun. Zheng.

a source of drinking water along the street, but it was also used for the decoration of buildings, squares, gardens, and monuments (Figure 1.5(a), (b)). Many were built as water jet fountains, and the number of fountains increased quickly to display the glories of the city, the memory of the victory of the war, and monuments to the aqueduct construction. In the medieval era, a center fountain was considered as the symbol of the city and the symbol of power. Fountains also played a religious role in history, and were characterized with religious meanings. They were also popularly used in Christian rituals and in Muslim life, for example, for cleansing before people go into the Mosque. Furthermore, it is used to display artistry by diversified forms and sculptures, which makes it an artistic highlight and symbolic of the city. Fountains were continually constructed in recent centuries with the broad meanings which the city would like to express such as the history, religion, symbolism, politics, decoration, recreation, arts, etc. At the same time, many water supply facilities were built mainly to convey water to fountains. Also, it was not just a final structure of water use but also a cultural phenomenon driven by water management from the past to the present.

1.3.3.3 The level of institution

It is also broadly demonstrated that huge social cultures were generated from water management such as social norms, social relations, organizational principles, social networks, rules, etc. In many places, social relations were structured depending on the water management and the irrigation system. Societies were organized by the irrigation situation, and social relationships were also formed accordingly. Typical cases are those of the Acequia in Valencia of Spain, the southwest USA, and eastern Mexico desert areas. Acequias or irrigations canals represent a water technology transmitted from Central Asia to the Iberian Peninsula in the 8th Century, where they were further developed before being transmitted to America in the 15th Century by Spain. Acequias were combined with the ditch technique and water management institutions to ensure that water could reach the target lands for irrigation. Local social relationships were shaped by irrigation management. A community was usually formed along a ditch, while some ditches formed a larger community with an irrigation community called a town. In Valencia, the leaders from eight major ditches formed the Valencia Water Tribute, which played the role of administration for irrigation. This has been functioning until today from AD 12. In New Mexico, the town leadership was the irrigation leadership. The irrigation community not only shaped the water management rules, norms and social network, but it also shaped local knowledge, lifestyle, customs, rituals, festivals, etc.

In China, it was a common phenomenon that a city, a town, or a village was formed by water management, and a local society was constructed depending on water management. For example, many major cities were formed along the Jing-Hang Grand Canal as a result of the canal construction and function. As the foundation of Chinese society, the village is a more intuitive view to understand how the society was formed by water management. Before 1950s, the social relationship of rural societies was broadly shaped and related to water management. For example, in the case of Jin Shui (晋水) River basin, there are around 30 villages formed along the basin. Jin Shui River is a historic river in Shanxi province, sourced from the spring Nan Lao. The irrigation history dated back 2000 years along the river, and there are four main irrigation ditches built historically for basin agriculture. Water was the main source of the basin's subsistence. The villages along the basin were formed by the river water resource management for agriculture. Therefore, this formed the social relationship, norms and rules for water management such as: (1) The village leadership was formed by the villagers depending on water management. Thus, the leader of the kin will become the leader of a ditch, called 'Ja Zhang', and will manage the ditch affairs. (2) There were rules to control water use for irrigation, and from upstream of the river to the downstream will employ a shift in using water daily according to the regulation. (3) The social norms include yearly ditch repair, mediation of the dispute, and impacts to manage the interests of the basin. (4) Organizing the yearly river god, 'water mother', worshipping ritual, at the temple of the river is a very important ceremony for the whole basin (Wu, 2011). The god worship is not just a spiritual activity, but it also played the role of consolidating the identity of the basin, maintaining the harmony of the basin society, and saving water as a kind of social institution (Na *et al.*, 2015).

To sum up, water management was important to drive the generation of culture. This is done from the spiritual level to social level, institutional level, and others in a comprehensive scale in human society.

1.4 CONCLUSION

In conclusion, water has always been comprehensively connected with the human society. Thus, this has resulted in diversified ways to think about, use, and manage water, and related institutions and techniques. Consequently, this has formed water culture, which is a sub-culture of humankind. This culture has influenced water management worldwide and it is characterized by ethnic, national, local, and religious backgrounds. Since the water cultural elements exist at the levels of ideas, social institutions, behavior and physical constructions, it has also formed a structural cultural system including mutually influential elements under the conditions for water management.

REFERENCES

Angelakis A. N. (2015). The history of fountains and relevant structures in crete, Hellas. *Journal for Global Environmental Issues*, **14**(3/4), 200–215, http://dx.doi.org/10.1504/IJGENVI.2015.071865 (accessed 5 May 2019).

Na L. X., Yan G. and Fu L. Y. (2015). Thinking on Shui Mu belief and sacrifice circle in the Jin River Basin of the Ming and Qing Dynasties. *Journal of Taiyuan University of Technology*, **33**(2), 39–42.

Ningbo News. (2014). Originality photos at a walk to Tuo Shan Yan Dam. Impressions of Yinjiang. *Ningbo News*. No.18. Ningbo.

Petri S., Juuti P. *et al.* (2013). History of fountains: legacies and lessons. *Water*, **5**.

Tua L. G. Z. (2011). Shan fair and its faith. *The Journal of Shanqiu Occupational Technologies College*, **3**, 72–73.

Wang M. (2017). An investigation of the status quo of the belief culture of Da Yu in east Asia. *Foreign Language Research in Northeast Asia*, **4**, 10–14.

Wu L. H. (2011). The Memory of Jin Shui: a Water Management Community's Construction Historically and Recently. China Social Press.

Xiao Yun Z. (2013a). The theory and prospect of water culture. *The Thinking*, **4**, 1–9.

Xiao Yun Z. (2013b). Safeguarding water cultural heritage against climate change: with case of Yunnan Province, China. *The Social Science Front*, **10**, 121–128.

Xiao Yun Z. (2013c). Shaping Beliefs, Identities and Institutions: The Role of Water Myths among Ethnic Groups in Yunnan, China. In: A History of Water, T. Tvedt and T. Oestigard (eds.), Vol. **1**, Series II, I.B.TAURIS, pp. 405–423.

Xiao Yun Z. and Min S. J. (2017). Tuo Shan Yan Dam and City Ningbo: exploring the historical logic of urban water management through a big city of China. *Journal of Earth Science and Engineering*, **7**(3), 127–135.

Chapter 2

The cultural dynamic of water management: Practical demonstration

Xiao Yun Zheng

ABSTRACT

Water culture has played an important dynamic role in water management historically, and it remains highly valuable for water management and water sustainability today. Water management was influenced by water related value, ethics, knowledge, norms, social behavior, institutions, etc. The role of water culture in water management could be played out from the local to the national scale. As well as using traditional water culture, it is valuable to encourage water culture dynamics in the water management process. Water culture can be used in policy making, capacity building, social awareness establishment, behavior fostering, engineering construction, etc. Simultaneously, it needs to stimulate more new water cultural elements to meet the demands of water management. This chapter aims to discuss the cultural dynamic mechanism in water management, practical demonstrations of water culture in current water management and how to foster new water culture for sustainable water management.

Key words: cultural dynamics, cultural factor, cultural inheritance, water management, current practice, sustainability

2.1 INTRODUCTION

Cultural factors are emphasized not only in the study of water: cultural influences on sustainability, environmental management, natural resource management, and water management are being increasingly emphasized. For example, culture is recognized

Editor: Xiao Yun Zheng
doi: 10.2166/9781789062045_0019

as one of the four pillars of sustainable development by the inter-governmental organizations and scholars (Blond, 2013; p. 15). In the field of environmental resource management, researchers studying the valuation and protection of ecosystem services now recognize cultural services as one of the most compelling reasons for conserving ecosystems. Therefore, cultural ecosystem services are highlighted. In reference to culture related to ecosystem management, some scholars suggest that it should be defined using four aspects: 1. Cultural worldviews and epistemes; 2. Cultural symbols; 3. Cultural assets; and 4. Cultural institutions, practices, or forms. Since the purpose of cultural ecosystem services is to capture the intangible attributes of nature, the focus is on ecosystems as generative of knowledge and supportive of human experiences (recreational, aesthetic, social, and spiritual) (Satterfield *et al.*, 2013). In the traditional approach to manage and sustain natural resources, culture is usually absent or is construed as a 'black box' and treated as an exogenous variable. For sustainability, to open up this 'black box', it was suggested that we should move away from the traditional way, which is the complex of biophysical and social-economic systems, and create a new way which includes culture. This is because of the crucial influence of culture as a part of an analytic framework (Caldas *et al.*, 2015). In reference to water management, which is an important aspect of natural resource management, apart from the natural flows of water, the concept of cultural flow for integrated water resource management has emphasized in recent years. Cultural flows refers to the dynamic ways in which water sustains diverse cultural beliefs, values, and ways of life, as well as the ways in which groups value, care for, and sustain the health of aquatic systems. This approach involves not only managing flows to sustain ecological processes, but also ensuring that the water flows allocated to non-consumptive or environmental purposes are sufficient to recognize, respect, and sustain the cultural ways of life (Johnston, 2012). The idea of cultural flows considers unnatural factors, non-consumptive factors, and the cultural factors, such as beliefs, worldview, norms, lifestyles, aesthetics, and recreation in water management. This expanded water management in an integrated way so that water is managed not only in a natural based way but also in a cultural way. There are benefits attached to understanding not only the cultural role for water but also the strategic and decision-making process. Accordingly, culture as an important factor in natural resource management has become increasingly recognized.

There are huge cultures which were generated by water management. At the same time, culture as a dynamic factor played a positive role in water management. This has been demonstrated historically and practically. Water management was influenced by the water related value, ethics, knowledge, norms, social behavior, institution, etc. Thus, this can also be surveyed from different levels of human culture. For better and sustainable water management, it is valuable to understanding the role of culture in water management historically and to renew the water culture in our contemporary situation.

2.2 THE SCALE OF CULTURAL DYNAMICS IN WATER MANAGEMENT

Water management is a broad idea from a micro level to macro level, so how wide is the role for culture in water management? Especially is it possible for culture to play a role in large scale water management nationally or globally?

Small scale water management is that which occurs at the local level, for example a village, a town, and a small river basin. There follow some case studies from China to discuss the cultural role in small scale water management.

In Di Qing Tibetan Autonomous Prefecture, Yunnan Province of China, the local ethnic people believe that the mountains and lakes are holy places. The lakes are believed to be the house of the dragon and other gods. As a result, almost all the lakes in the local society have secret meanings. Local people believe that the dragon and the gods of the lakes controlled the rain and the peace of human beings. Therefore, if people respect the gods of the lake, they will receive peace, personal safety, health, and happiness. On the other hand, the gods will bring flood, illness, and other disasters to humankind if they are disrespected. Accordingly, the lakes have been safeguarded carefully by the local people. There is no hunting, felling of trees or pollution (bathing, urinating, etc.) close to the lakes, and even no noise at the lakeside. Also, local people perform worshipping rituals to the lake gods every year. This belief was valued for protection of the lakes until recently and it is even more powerful than the official rules and law. Therefore, the uncontaminated state of of the lake water has been protected by traditional ideas and practices (Wang, 2005).

Dai is an ethnic nationality in Yunnan province of China. Dai people believe that the universe originated from water. The idea of water in Dai society is that water is a substance which made the universe, sustains human life, and controls subsistence. At the spiritual level, water is a substance that could purify the soul. Also, in their Buddhist faith, water is a medium between the holy world and material world. At the material level, water is not just an indispensable substance for life but people rely on it for subsistence, and all subsistence started from water. Therefore, water was strongly respected in Dai society. The idea of water also influenced practical water management in Dai society. As discussed earlier, water shapes knowledge, systems of nature, and production. For example, a popular saying expressing the cycle of water and subsistence is: 'If there is water, the forest will keep growing. If there is water, the paddy field will be cultivated. If there is a paddy field, human life will be continued.' People believe that water is the mother of forest, and forest is the home of water. As a result, people have to protect both water and forest. Before the 1950s, there was 90% forest cover in the Dai people's homeland which was regarded as the secret forest and protected by secret meanings. Hence, people could not hunt, collect, or fell trees, or graze livestock in the forest. A healthy forest would ensure enough water supplies to the rivers and ditches that supported the large scale irrigation system for wet-paddy cultivation. At the same time, rich water related customs, norms, and rules had also played a significant role in restricting the people's behaviors towards the water facilities and helped protect the irrigation channels. For example, customs restricted people to bathing and washing near the village drinking well. The water source and forest were strictly protected against entry under the secret meanings and rules. The norms and rules required the villagers to protect the irrigation channels, cause no harm to the channel's function, take reparation yearly, and distribute water equally. Subsequently, the water related culture has played an important role in water management in Dai society from daily life to their productive activities.

In summary, the water culture acts positively in local water management. The local society relies on culture to manage water in an integrated, effective, and sustainable approach.

Culture has not just played an important role in water management at the local scale, but also on a large scale both historically and currently. For example, in the Roman Empire, many grand hydroengineering projects were constructed for urban water supply. The urban water system which usually consisted of an aqueduct, tank, distribution facilities, and water consumption system. The water consumption system mainly consisted of the bath, toilet, fountain, etc. The technological and constructional achievements of imperial Rome represented the historic peak of water management until the industrial revolution in Great Britain in the 19th Century (Bell, 2009; p. 47). Consequently, the cultural meanings in the Roman Imperial era consist, firstly, of the idea of water management. Large scale water management was not just to meet the demands of urban water supply, but it was also a symbol of imperial power and a glorious landmark to show the transcendence beyond the preceding Greek civilization. The Roman aqueduct and matched facilities were built in imperial territory. At the same time, sufficient public water supply was a source of the legitimacy of the empire. Secondly, the lifestyle and customs was typically reflected in bathing. Bathing was an important facet of lifestyle and a cultural way to reach the unity of the empire. Thus, baths were built in the imperial territory. In many cities, the bathing facilities were the first public constructions, and some aqueducts were built only to supply water to the baths (Hodge, 2011; pp. 5–6). The Roman bath remains in today's Rome can be visited, as well as those in other ancient Roman cities, to understand the importance of the baths. For example, in the Roman city of Aptera in Crete, the public baths were the outstanding construction of the city and their ruins still endure. One of the most important public constructions was two large scale cisterns that functioned mainly to supply water to the baths. Since Aptera is located at the top of the hill, the only possible water supply was rainwater. The precious water was mainly used for bathing which reflected the importance of bathing in social life, which was driven by cultural idea, lifestyle, and customs. The importance of the baths was not the only influence on water management in the Roman Empire, but there were many others. For example, fountain, including public drinking fountains, monumental fountain and decorative fountains, were built in large numbers in the Roman cites. This was driven institutionally and aesthetically and also influenced water management. In essence, water management was determined by cultural ideas, lifestyle, and social norms which were common in the Roman Empire.

2.3 WATER CULTURE AND WATER MANAGEMENT

Water culture exists as a series of elements and levels; they are intertwined together as well as functioning separately in water management.

2.3.1 Cultural idea and intentions

An idea is a positive factor for people that causes them to think about water and to protect water. Ideas about water can arise from religion, world view, or practices of water management. The idea of water has influenced the human behavior towards water. The idea of water in religion is also an example of the cultural phenomenon of the 'Given factor' (Veronca, 2004). This means there are many secret and holy meanings of water which came from religions. For example, it is a popular phenomenon that the Chinese ascribe religious meanings to springs. Thus all springs have holy meanings and this led to the establishment of the influential 'dragon springs' cultural system and consequently, the springs were protected. Many waterscapes were destroyed historically but the springs were protected and are in good condition in China and thought of as holy places. Spring water is also protected, and springs are usually important water sources for local for irrigation and urban water supply. Related cases are easily found in China. Therefore, the religious ideas on water management can be surveyed against a broad background. For example, water management in Muslim societies has particular characteristics. Water fountains are often built close to mosques so that people can wash before entering the mosque. Islamic decorated fountains and garden waterscapes were made in a special way, with corresponding water management. Water management not only depends on religious idea but also on other intentions. For example, it may depend on the idea of Feng Shui, which is an ideal and practical system of humans and nature in China. It is also a common phenomenon that the Chinese included waterscapes in the residential designs to express wishes and make the residents more comfortable. In the central provinces, such as Zhejiang, a large pool was usually constructed as a part of the village. It expressed a cultural idea but was also important for water supply in the locality. Accordingly, the pool and matched water system have been protected until today in most cases.

2.3.2 Aesthetic demand and leisure

Water is a pleasant substance. For this reason, water management to meet aesthetic demands has been practiced in human society from ancient times until today. For example, many waterscapes are typically constructed to meet aesthetic demands and for leisure pursuits. This happened not just in urban areas, but the countryside and village also depended on rivers, lakes, and springs built artificially. Such cases are easily found worldwide, for example Venice (Italy) and Beihai Lake in Beijing (China). In order to maintain the waterscapes, water has to be managed in a particular way either through the management for water supply or by preserving nature. Today, more and more waterscapes have become scenic places. Other cases like water facilities of garden construction and fountain also reflected aesthetic demands and require particular water management. In Europe, some fountains had a dedicated water supply and the aqueduct made it possible to build fountains. For example in Perugia (Italy), the central fountain had its own water supply from an aqueduct.

2.3.3 Lifestyle

People adapt to local environmental condition, including water conditions, to shape their lifestyles. Usually, people have to change the local water condition to meet their demands, safety, and comfort, and maintain their lifestyle. Facets of lifestyle related to water include aesthetics and its practice, the custom of daily water use, water related residence, transportation, and social activities. Also included were festivals, rituals, games, recreation, and personal daily life.

Many towns and cities were constructed along rivers. In many cases, a city encompassed a series of urban rivers in China. For example, in Shao Xin City, Zhejiang province, there were 23 urban rivers inside the city before the 1960s. This means that the local lifestyle was shaped by the water situation. The residential houses were built along the rivers preferentially, and the local lifestyle was compatible with this situation. The daily water use, drainage, washing, transportation, and subsistence strongly relied on the urban rivers. The lifestyle was also locally shaped. In order to maintain the local lifestyle, people took measures to keep the water in good condition. This has been important work for the local governments and the residents for many generations. However, the extent to which the lifestyle is influenced by water depends on local water conditions. Nonetheless, water is an important factor that constructs and maintains human lifestyles.

2.3.4 Social relations

The water related social relations which include organization, networks, and social norms are usually generated from water management, and have played an important role in water management. Sharing, controlling, and using water usually depend on social relations, and this is possible for a community, an organization, or a larger network. For example, in Valencia (Spain), water for irrigation has been managed by the local Acequia societies, according to their social relations, local rules, and norms. This has been supervised by the Water Tribunal for almost 1000 years until today. In China, there are many cases which show that water was managed by local societies, depending on the relationships of the families, as well as the rules and norm which are agreed upon among the family-based stakeholders. Water management could build a network among the stakeholders. However, cooperation is required among the stakeholders to make the principles, rules, and norms for water management. For example, in the Yuanyang area, in the midstream valley of the Red River, 100,000 ditches were built to convey water from the top of the mountain to irrigate the terraced paddy fields on the hillside. Some of the ditches are 40 or 50 km in length and flow across a series of villages, from upstream to downstream. According to the ditch construction, a ditch-based social network was shaped along the ditch. At the same time, it also shaped the water related interests in relation to different sections. The inhabitants of the upstream villages are mainly interested in maintaining the water resources and ensuring water supply to the ditches. The midstream and downstream villages are the beneficiaries to varying degrees. The major concern for all the villages along the ditches is to ensure a safe water supply, a smooth flow, and that water supply is shared equally in order to meet irrigation demands. How a ditch can be managed in a good condition depends on the following circumstances. Firstly, an interest-related community was established along a ditch. Therefore, this brought about the cooperation for water management, and the responsibilities and interests are shared in different sections. A committee of the leaders of the villages meets together yearly to discuss the maintenance of the ditch, water management, and other concerns. Secondly, a graded price from the downstream and midstream villages was paid to the upstream villages, and all the villages benefited. Thirdly, rules, customs, and norms are formed for water management, ditch repair, redistribution of water, punishing illegal water use, etc. Under this social environment, hundreds of thousands of ditches in the valley are maintained in a good situation. According to the practicalities of ditch operation, a ditch's operation in a good condition mainly depends on the effectiveness of social relationships (Xiao Yun, 2005a). Therefore, social relations cannot be absent in water management. This is because effective water management is sometimes difficult.

2.3.5 The social behavior for water management

Here, social behavior is understood as the behavioral pattern which individuals followed in a society depending on social values or ethics to manage water. It is a collective behavior followed by individuals, but it can be described individually as personal behavior. Therefore, the behavior should be defined at two levels: (a) the behavior might be shaped based on the social values, ethics, or other cultural ideas or (b) behavior might be shaped based on personal life experiences or other matter such as personal economic concerns. Both levels of behavior may influence water management. The level of the behavior based on social values should be defined as social behavior. It is valuable to understand how social behaviors have significant impacts on water management. Depending on values and ethics, people regulate their behavior to reach their aims for water management, such as using water, saving water, and maintaining water in good conditions. This includes maintaining the amount and quality water. Consequently, the significance of water management is usually determined by the social behaviors. In the context of current water shortages, water saving is an important commitment of water management. Research on the social, attitudinal, and behavioral composition of water saving activities using a sample of 1600 households from Devon (UK) was conducted to investigate saving water in the home. It showed that behavior is an important factor for saving water in the home. Recommended water saving activities include turning off the tap when cleaning the teeth or waiting until there is a full load before using a washing machine and these can help individuals to make considerable water savings, alongside having a water meter installed and using a water saving device in a toilet cistern. These behaviors may directly be influenced by factors such as the habits, economy, social desirability, environmental threat, and environmental concerns (Gilg & Barr, 2006). Therefore, behaviors play an important role in saving water in the home. Water saving also depends on the habits and life experience, as well as the values of environment actions. Based on environmental values, implementing ethics to shape behavior for saving water is encouraged, and it could form a new culture for saving water.

It has been demonstrated that social behavior has a significant impact on water management depending on its values and ethics. The values are driven by behavior to achieve different results in water management. It is not difficult to find cases worldwide to understand how people modify natural water to develop agriculture, industry, and residential areas. This has obviously changed natural water and has had results counter to human interests in sustainability. A paradigm was the case of Shaoxi City in Zhejaing Province, China. Seventeen urban rivers were landfilled to allow construction of urban roads in 1952–1977 under the incorrect value choice. At the same time, there were many water facilities and waterscapes, which were destroyed by the inhabitants because there were no water-friendly values to direct their social behavior at that time. This resulted in changes to the urban water system and difficulty of drainage, as well as the local lifestyle (Yan, 2012; p. 76). Accordingly, good water management and good water situation requires the water-friendly social behavior. If the social behavior benefits water, water will be managed in a good way, and such water management will result in human benefits.

2.4 THE INHERITANCE OF WATER CULTURE AND WATER SUSTAINABLE MANAGEMENT

As discussed above, culture has played an important role in water management historically. Currently, it is still playing a significant role. The historical water culture and its dynamic role is a way to understand good water management and think about effective and sustainable water management in the present time. This is because many principles of water management have not been generated in the present but in the society of four or five thousand years ago. Furthermore, this includes not only techniques but also the principles and knowledge which are transmitted from generation to generation and from the ancient times to the present age (Angelakis *et al.*, 2012). It is also a fact that water culture has formed a part of the system of human knowledge, values, lifestyles, customs, and behavior. Accordingly, water culture is an important component of human culture which was inherited from the past and renewed in the current age. Presently, many elements of water culture are lost or being lost in the present circumstances due to the changed natural environment, urbanization, water supply pattern, lifestyle, etc. Therefore, how to inherit historical water culture has become a challenge in the present age.

Firstly, a great deal of historical water culture is still composed of knowledge, values and life systems, and practices in current water management. For example, the water system constructions in ancient Greece and Roman have finished their functions, and most of them are ruins today. However, the principles of water management generated from those times still have a significant impact on current water management such as urban centralized water supply, long distance water transfer, water-driven toilets, and bathing room. Accordingly, it is impossible to cut the connection with the past in regard to current water management. It is valuable to inherit the historical wisdoms and principles of water management in current practice.

Secondly, history has shown that good water management is always cultural-based. Accordingly, good water management should be driven by a beneficial culture which not only depends on nature but also on local social situations. Water culture was generated in different ages based on the nature and social reality at that time. The historical water culture principles can be applied in the current age, but it is difficult to meet all the demands of current water management. Now water management requires a culture which is matched with the current water and social situation to meet the demands of water supply, water quality, and water sustainability. This involves constructing a contemporary water culture system and using it as the dynamics of water management.

The major challenge is the composition of current water culture, how to construct it, and how to make it become the dynamics of water management. I suggest that the modern water culture should be composed of historical and traditional cultural elements as well as the cultural elements which are based on water situation, security, quality, equality, landscaping and sustainability, etc. The water situation of the present has changed a lot, for example, fast urbanization has changed the rural water situation. One should be aware that giving up traditional water culture could result in the collapse of environmental knowledge systems, including water knowledge in communities. This could lead to deficiency of the values, norms, and cooperation for water management in urbanized areas. Current environmental problems usually result from cultural deficiency in local society (Xiao Yun, 2012). Accordingly, the best way to construct the current water culture should be based on historical and traditional water cultural elements as well as the demands of current water management. For example, talking about the water ethics, David Groenfeldt suggested that introducing a 'new water ethics' is now necessary, but this introduction needs to address the presence of the old water ethics, which have evolved within society, the economy, and the culture (Groenfeldt, 2013a).

However, it is a fact that the historical and traditional water culture is not enough to meet the demands of current water management because the current water situation is usually not a repetition of the past. Construction of a new water culture is needed and to help understand this some elements of water culture are discussed in the following sections.

2.4.1 New water values

Based on the ideal level of water culture, new water values are needed to achieve water management and water sustainability. The 'new' water values should not deviate from the past but should be more focused on the current water situation such as water shortages, pollution, quality, and equality. Also, consensus is needed about water management in terms of safety, equality, human rights, transparency, and sustainability. Here, the consensus must go beyond the traditional local society to a larger, even global, society. Today's elements should be formed to direct behavior in regard to water management and be directed by the common interests. Water value does not only depend on the local water situation and traditions, but it is also possible for it to become a global value for global water management. For example, water ethics not only concerns the community and indigenous people but also the global audience: for instance, the principles, policies, stratagems, and proposals from the UN and other international institutions depend on the common interests of humankind that formed the global water ethics as the frame of water management (Groenfeldt, 2013b). Nevertheless, the ideas, ethics, and identity depend on human common interests and are important components of new water values. Many of them are different from the past, such as water equality and sustainable water use.

2.4.2 Water saving

Water saving has been practiced with a limited awareness in many countries over the past half century or so. Increasing water supply was the main way to meet greater water demands, e.g., by building more dams and other facilities. It is impossible to continue to increase water supply without limit through engineering approaches. This is aimed at meeting the huge increased water demand following fast urbanization and industrial development due to the imbalance between nature and human demands, as well as against the background of climate change nowadays.

Therefore, water saving is now recognized as an important approach to coping with the water crisis. Water saving requires that water use be effective, economic, reasonable, and even ethical. Hence, it requires integration from awareness to behavior, institutions, and techniques. People also need to increase their awareness about the importance of water saving. Based on awareness and values, people also develop behavior to save water in their daily life and industrial activities. This further creates the measures, approaches, customs, norms, and institutions to save water. The combination of awareness, values, behavior, customs, and norms comprises a cultural body of water saving. It also brings water saving as a social movement as well as a cultural phenomenon. It can be regarded as the 'Culture of Water Saving' which can also be understood as a new culture of water.

2.4.3 Culture for water education

Currently, in order to cope with the water crisis, it is important to reach the goals of water security and water sustainability, increase the public awareness, and build public capacity. The key segments are: awareness and values that the public needs, how to bring them to the public, and how to increase the capacity of the public to solve the water problems. Without a shadow of doubt, education plays a key role. The culture for water education depends on educational ideas, the importance of education, and how to practice education.

Firstly, it is necessary to retain historical and traditional knowledge as well as to increase awareness of security and knowledge of water use by means of friendly and sustainable approaches. It is also important to build values which match with the current intention of water management. Traditional ideas are very helpful to support the building of current values of water use. For example, traditional ideas of water are related to animism and Hinayana Buddhism, and the idea of cycles of water, nature, and human life effectively support water protection in the Dai ethnic society in China. These ideas are taught locally by the approaches of school, temple, and by public ways (Xiao Yun, 2005b). The Tibetans in Yunnan province of China also considered water as a very secret substance according to their Buddhist faith. Thus, their behavior to water is strictly limited by their faith, and it is also very helpful to water protection (Zhang Shi, 2011). Accordingly, water education should be based on the common demands of coping with the current water crisis.

Secondly, the practice of water education may involve diversified approaches but focuses on a consistent goal. For example, water education, as a part of school courses, is popular in many communities in Japan, Thailand, and elsewhere. Currently, many innovations institutionally and technologically also increase the effectiveness of water education. A typical case is the establishment of Global Network of Water Museums – an initiative endorsed in 2018 by a dedicated UNESCO Resolution. This is an additional confirmation of the imaginative potentiality of innovative research approaches aimed at bridging, in a holistic perspective, past and present water civilizations with their related heritage and management models. Such a research perspective is fundamental to inspire new approaches towards water sustainability education and water awareness efforts in the frame of implementing the UN's targets of Sustainable Development Goals. Nowadays, museums are engaged all over the globe in implementing water awareness and educational activities which emphasizes new possible links among past and present water civilizations, that is, among the inherited water culture and the more recently integrated water management models aimed at strengthening sustainability (Eulisse, 2021). The Global Network of Water Museums is an innovative means of water education, involving new ideas, institutions, and practices. In addition, other water educational innovations such as digital museums, open hydro-work places, for example, the Hydro-scenic Park in China with the functions of recreation as well as education, have formed the new approach of water education. Consequently, the ideas, contents, ways, and practices of water education have formed a 'new water culture' for water management.

2.4.4 The way to stay with water naturally

Globally, this is the era of fast urbanization. Before urbanization, human beings had a close connection with nature and were closely associated with the natural water environment. However, after urbanization, people became distanced from water in nature and the natural state of water was transformed. This included creating waterscapes and rivers for the construction of dam and city, and also the use of water by current water supply system. This situation might harm nature as well as the sustainable interests of human beings. Accordingly, how to stay close to nature, including water, should be an important benefit for human beings. Therefore, water should be kept in as natural situation as possible during urbanization and construction. Also, attempts should be made to return water into a natural state; to make the river flows freely, the artificial river banks should be re-profiled; the waterscape should balance with human construction, and the exploitation of water resource be reduced. To meet these objectives, a new culture should be formed which includes values, ethics, lifestyle, behavior, institutions, etc.

In summary, the traditional water culture which is inherited from history is still valuable and many elements are functioning nowadays. At the same time, new elements of water culture are needed to meet the current water situation and to solve water problems. Accordingly, many new ideas, values, lifestyle, social norms, social behavior, and technologies and mechanisms of innovation should become part of the water culture for water sustainable management.

2.5 THE DYNAMIC ROLE OF WATER CULTURE IN CURRENT WATER MANAGEMENT

How water culture plays a dynamic role in current water management is the key to displaying its value. As discussed above, water culture has played an important dynamic role in water management in the past, and culture is also needed for current water management. Water culture should not only play the traditional role, but it also needs to meet current demands of water management locally and globally. Therefore, the current dynamic role of water culture should have many differences compared with traditional role in the past. The dynamic role water culture plays is seen below :

2.5.1 The dynamic role in policy making

Water policy determines how water is managed. The policy making should be based on the real demands, for example, to meet the demands of industrial development and urbanization. However, it also has to be based on culture, since it is an unfinished issue. Good policy making should consider the water related value, ethics, tradition, custom, local participation, etc. Consequently, this makes the water policy making more effective and sustainable. For example, equal value determines how the water policy is made in a community to avoid possible water impacts and to make water management in a sustainable way. If the local water policy making considers the tradition, local concerns, and participation of the community, it will be more acceptable to the local society for implementation of policy. It is not a new idea, but it is truly important to understand the role of water culture. Unfortunately, although important cases show that cultural absence in water policy making is still a reality, water culture can play a dynamic role to direct the water policy making it equal, transparent, acceptable, and sustainable.

2.5.2 The dynamic role in hydro-projects

Currently, huge hydro-construction is on the way in many countries, especially construction of dams. At the same time, it is frequently reported that dam construction brought many impacts to the local societies, and some dams have even been stopped by local people. There are many reasons for these impacts, but one is the absence of the cultural concerns in the project planning and practice; in other words local participation, values, traditional lifestyles, ways of subsistence, and social relations are usually neglected. Therefore, an effective and sustainable hydro-project planning and construction should be more culturally inclined by making culture a part of the planning and construction. The cultured hydro-project should include the purpose of construction as well as the local values, lifestyles, and traditional interests to make the project sustainable.

2.5.3 Water culture as the social foundation of water management

Cultural-based water management will make water management more effective and sustainable. The value, awareness, water friendly behaviors and norms, and institutions are the foundation of good water management. In order for water management to be effective and sustainable, there is need for cultural support.

This would be a common principle. An environment-friendly society should reduce the cost of water management, reduce water pollution, and increase the effectiveness of water saving. Accordingly, water culture should support the current water management. Fundamentally, traditional water management is usually based on life and industrial demands as well as local cultures, but the purpose of current water management usually goes beyond the local demands and traditions. It may, however, be directed by common values and interest of mankind, for example, saving water, water equality, water security, etc. Accordingly, a culture which includes the elements in at the local level as well as the elements at the global level is necessary for water management. For example, the values of equality, transparency, security, human rights, and sustainability foster water friendly behavior, social relationship of cooperation, and establish the good custom for water saving. Making culture the foundation of water management will greatly benefit the rise of the effectiveness of water management to achieve sustainable goals locally and globally.

2.5.4 The role of water education

To achieve the goals and purposes of water management, inheritance of water traditions, increasing knowledge and awareness, fostering values, building capacity, and friendly behavior are important. As a part of water culture, water education could play an important role because water management will be ineffective or fail without social support. However, social support is not naturally formed in a society but develops from life experience and practice of water management, either inherited or taught. If values, ethics, and social behaviors are not fostered, it will be difficult to reach the goal of urban water saving and prevent water pollution. People would not understand why it is important to save water especially in public places where the water use is not based on personal payment. Moreover, some people throw garbage freely into the rivers in some cities, and this becomes a big problem for local river water management. Thus it has been suggested that the absence of environmental ethics for some people living along the river is the main reason for their actions. Research strongly suggests that education is the most important way to raise environmental awareness and ethics of the public to stop this pollution. Therefore, education is an indispensable way to foster environmental ethics for the local residents (Xiao Yun, 2020). Education can support water management and strengthen the social foundation of water management. However, it is also a fact that water education is not a normal part of national education in most countries so not all students will be educated in terms of water knowledge and values. Alternatively, water education may be implemented in museums, temples, and NGOs which forms a specialized education system locally and globally. Depending on the common value and goal for water management, water education displays diversified ideas, contents, approaches, and measures, both locally and globally.

It should be noted that culture is not always accepted as part of water management. This is because cultural-based water management may increase the social and financial costs situationally, thus, water ethics may go against economic interests in a short time, and this could limit use of water resource in a sustainable way. More social coordination and consultation with the local society will be needed to make engineering construction, such as dams, culturally based. Culturally aware water policy making requires more round-table cost compared to the single criterion of economic purpose. However, paying these costs is valuable in helping to achieve sustainable water management. The effectiveness of culture and the dynamic role it plays in water management should be increased by education, transmission, dialogue and innovative institutions, mechanisms, and techniques.

2.6 CONCLUSION

Water culture is generated from water management and other purposes. People worldwide have always relied on their ideas, values, religion, social relationships, and institutions to manage water, especially in non-urbanized times and areas. Therefore, water culture has played an important role in water management, and this has made water management acceptable, effective, equal, and sustainable. Unfortunately, however, water culture has been often neglected in current water management which is driven by urbanization, marketization, and social system transformation. There have been clear impacts on integrity, efficiency, and sustainability in water management. Accordingly, it is necessary to recall the importance of water culture. Water culture has been inherited from the past, and it is also renewed in the present age. Currently, it needs to generate new water cultural elements to meet the demands of water management. This is because the water situation has changed a lot compared to the past. It is valuable to inherit the traditional water culture as well as to encourage the water culture dynamic in the water management process, for example, in policy making, capacity building, social awareness establishment, behavior fostering, engineering construction, etc. Accordingly, water management should not only be a technical process but also a cultural process. This is seen as a better way of sustainable water management.

REFERENCES

Angelakis A. N., Mays L. W., De Feo G., Salgot M., Laureano P. and Paranychianakis N. (2012). Water and wastewater in ancient civilizations. In: A Compendium of Hot Topics and Features from IWA-SG on Water and Wastewater in Ancient Civilizations (WWAC). Global Trends and Challenges in Water Science, Research and Management, H. Li, (ed.), IWA, London, UK, pp. 90–94.

Bell A. (2009). Peak Water: Civilisation and the World's Water Crisis. Luath Press Ltd, Edinburgh.

Blond O. (2013). Culture is the fourth pillar of sustainable development. In: Water Heritage: Material, Conceptual and Connections, Willem J.H. Willems and Henk P. J. van Schaik (eds), Sidestone Press, Leiden.

Caldas M. M., Sanderson M. R., Mather M., Daniels M. D., Bergtold J. S., Aistrup J., Heier Stamm J. L., Haukos D., Douglas-Mankin K., Sheshukov A. Y. and Lopez-Carr D. (2015). Opinion: endogenizing culture in sustainability science research and policy. *Proceedings of the National Academy of Sciences*, **112**, 8157–8159.

Eulisse E. (2021). Unveiling Venice's Waterways Heritage. From the Digital and Open-Air Water Museum of Venice to UNESCO's Global Network of Water Museums, this volume, pp. 00–00.

Gilg A. and Barr S. (2006). Behavioural attitudes towards water saving? Evidence from a study of environmental actions, *Ecological Economics* **57**, 400–414.

Groenfeldt D. (2013a). Water Ethics: a Values Approach to Solving the Water Crisis. Routledge, Taylor & Francis Group, p. 156.

Groenfeldt D. (2013b). Water Ethics: a Values Approach to Solving the Water Crisis. Routledge, Taylor & Francis Group, p. 157.

Hodge T. A. (2011). Roman Aqueduct & Water Supply. Bristol Classical Press, London.

Johnston B. R. (2012). Water, Cultural Diversity, and Global Environmental Change: Emerging Trends, Sustainable Futures? UNESCO Publishing and Springer, p. 15.

Satterfield T., Gregory R., Klain S., Roberts M. and Chan K. M. (2013). Culture, intangibles and metrics in environmental management. *Journal of Environmental Management*, **117**, 103–104.

Veronica Strang (2004). The Meaning of Water, Berg, Oxford, pp. 83–102.

Wang X. S. (2005). The Holy Mountain Worshipping and Ecological Protection, in Tibetan Culture and Biodiversity Conservation. Yunnan Science & Technology Press, pp. 79–82.

Xiao Yun Z. (2005a). Ethnic water civilization and irrigation in the Chinese reaches of the Red River. *Forming Japan*, **39**(5), 35–41.

Xiao Yun Z. (2005b). The water culture in Dai society and the sustainable development. *Thinking*, **31**(6), 33–39.

Xiao Yun Z. (2012). Contemporary Environmental Issues and Their Cultural Backgrounds: Anthropological Analysis on Some Cases of Yunnan. In: Current Problems of Environment and its Cultural, X. Zheng (ed.), China Book Press, p. 23.

Xiao Yun Z. (2020). The role of social sciences in coping with the current water crisis. *Journal of Tsinghua University*, **35**, 177–188.

Yan Y. (2012). The Water Culture of Zhejaing River Way Engineering: Shaoxi Section, Center for Water Culture Research and Education, Zhejiang University of Water Resource and Electric Power, *Zhejiang*.

Zhang S. (2011). Water culture of the Tibetans in Diqing of Yunnan. *Journal of Yunnan Normal University*, **43**, 43–48.

Section 2

The Historical Dynamic Role of the Culture

Chapter 3

Urban hydro-technologies in Crete, Greece through the centuries

Andreas N. Angelakis

ABSTRACT

The history of water supply and wastewater engineering in Crete, Greece dates back more than ca. 5,000 years, to the early Bronze Age. One of most advanced achievements in Minoan palaces and other settlements was the architectural and hydraulic functions of water supply and sewerage systems. Minoan hydro-technologies were diffused to Greek mainland, where more advanced sophisticated hydro-technologies were invented, with a peak during the Classical and Hellenistic periods, when they spread over from Greece to India in east and Egypt in the south. The Roman Empire, which replaced Greek rule in most of this area, inherited the hydro-technologies and further developed them, thereby changing their scale of application. Urban water and waste- and storm-water management in Crete were very little improved in both in principle and in scale, during the Byzantine and Venetian periods; however, those constitute the underpinning of modern achievements in water engineering and management practices. The scope of this chapter is to present the most characteristic examples of ancient hydraulic works and related hydro-technologies and their uses through Cretan civilizations.

Keywords: Aqueducts, cisterns, drainage and sewerage systems, fountains, wells, water distribution, prehistoric and historical times, modern times

Editor: Xiao Yun Zheng
doi: 10.2166/9781789062045_0037

3.1 PROLEGOMENA

3.1.1 The past is the key to the future

Crete was the center of Europe's first advanced civilization, the Minoan (Mays *et al.*, 2007). The first human settlements on the island date back to the aceramic Neolithic period; ancient Knossos was one of these major Neolithic (then later Minoan) sites. Since the Neolithic Age, Crete's climate has remained more or less unchanged (Markonis *et al.*, 2016). Nowadays, the climate is mild with relatively warm winters and hot summers. The mean annual temperature ranges from 17°C to 20°C. In general, the Mediterranean climatic zone influences a large part of Crete and makes spring and autumn short in comparison to the longer summer and winter seasons. Crete, following its initial homogeneity in early antiquity and its gradual development since the Minoan Era to a growing multicultural centre, fell to various conquerors throughout its long history. In the course of time, urban water supply to its various settlements remained a vital issue (Tsonis *et al.*, 2010).

The oldest evidence of modern human habitation on Crete is pre-ceramic Neolithic farming-community remains which date to about ca 7000 BC (Broodbank & Strasser, 1991). A comparative study of DNA haplogroups of modern Cretan men showed that a male founder group, from Anatolia or the Levant, is shared with the Greeks (King *et al.*, 2008). The island of Crete was first inhabited shortly after the late Neolithic period, but it was only during the Bronze Age that the Minoan civilization developed and reached its pinnacle (Alexiou, 1964). The Minoan and Mycenaean settlements, mainly in Crete and Peloponnesus, respectively, developed and applied various technologies for collecting, storing, transporting and using surface water and groundwater resources (Angelakis, 2016; Angelakis & Spyridakis, 2010; Koutsoyiannis *et al.*, 2008). At that time the island was probably divided into four political units, the north being governed from Knossos, the south from Phaistos, the central eastern part from Malia and the eastern from Zakros (Angelakis & Vavoula, 2012). Smaller settlements have been found in other places (Mays, 2007).

The progress in urban water supply and waste- and stormwater was even more noteworthy, as witnessed by several aqueducts, cisterns, wells and other water facilities discovered, including the famous Minoan aqueducts of Knossos, Malia and Tylissos, the cisterns of Zakros, Archanes, Myrtos-Pyrgos and Tylissos, the wells of Palaikastro, Zakros and Itanos (e.g., Angelakis, 2016, 2017a; Koutsoyiannis *et al.*, 2008). Beginning in 1420 BC, the Minoan civilization was overrun by the Mycenean civilization from mainland Greece. Thus, the advanced Minoan water technologies spread to the Greek mainland (Angelakis & Spyridakis, 1996). In the Archaic and Classical periods of Greek civilization, aqueducts, cisterns and wells similar in principle to the Minoan and Mycenaean originals were constructed (Angelakis & Vavoula, 2012). The sewerage and drainage techniques in ancient Athens are almost the same in principle as those applied in Crete and the Indus valley about 4,000 years ago and remained largely unchanged until the Middle Ages (Wilson, 2000). In Roman period, the scale of the projects was largely increased (e.g., the Cloaca Maxima in Rome). Public toilets facilities became common and private toilets usually lacked running water (and were commonly located near the kitchens). Dysentery and other kinds of diarrhoeal diseases were common (Angelakis & Rose, 2014).

From about 330 to 1450 AD, traditionally referring to the Middle Ages or Byzantine period, when a demographic, cultural and economic deterioration occurred in western Europe and of course in Crete following the decline of the Roman Empire, very little progress was made. A new concern with health and water issues arose during the Renaissance (12th–17th century) (Koutsoyiannis *et al.*, 2008).

Finally, at the end of the 19th century and the early 20th century, after the intense industrialization and urbanization, the first modern urban wastewater treatment plants began to be constructed in Central Europe, the United States and other parts of the world (Angelakis, 2017a). Also the first modern water distribution systems appeared in urban areas. In addition, modern wastewater treatment technology began to have a significant impact on improving the quality of drinking water, protecting the urban environment and hygiene conditions (Angelakis, 2017b).

In the present study, an attempt is made to trace the history of urban hydro-technologies in Crete's major historical phases from the Minoan times to the present.

3.2 MINOAN CIVILIZATION (*CA.* 3200–1100 BC)

Evidence for the first human settlements in Crete goes back to Neolithic period. Knossos was the center of the Minoan civilization (*ca.* 3200–1100 BC), the first and most important civilization in Europe and the Mediterranean. The Minoans spread over Crete and the islands of the south Aegean Sea (e.g., Santorini). Archaeological evidence indicates that besides the know-how relevant to urban planning, construction and ceramics, the hydraulic and water-related and sewerage sectors were well developed. Later, scientific fields, such as hydrogeology, aqueducts, hygiene, baths, water and sewage networks, and the use of water for recreation were established. The remnants of various relevant structures suggest that the Minoan water resources engineers had a very good level of knowledge of basic concepts and principles of water and wastewater management in a sustainable way, allowing the Minoans to live in harmony with their nature and environment (Angelakis *et al.*, 2012, pp. 227–258).

Villages, towns or palaces were often founded on isolated hills and mountains as these sites had strategic advantages when defending the settlements. Water supply

Figure 3.1 (a) Part of the aqueduct in the entrance of Tylissos villa and (b) rectangular terracotta pipes used for rainwater collection and storage.

for human beings and animals at these settlements had to be secured, especially because of the possibility of being besieged for long periods of time. Examples of Minoan achievements in water technologies have been reported by Angelakis *et al.* (2005, 2012, pp. 227–258) and Koutsoyiannis *et al.* (2008). Minoans gave careful thought to water supply and by developed three major technologies: wells to reach groundwater, cisterns to store rainfall and aqueducts to transfer water from springs or surface sources (Antoniou *et al.*, 2014). Two examples, the end of the aqueduct of Tylissos village and in Pyrgos-Korifi, Myrtos in southeastern Crete, are shown in Figure 3.1.

Minoans had also developed sanitation technologies (e.g., waste- and storm-water sewers, baths and toilets. Several Minoan cities and palaces had well-established sewerage and drainage systems, which are in good functional condition even today (Angelakis & Rose, 2014). In the entire history of humankind, there exists no other example of a sewerage and drainage system still functional more than 4,000 years after its original construction (Figure 3.2a). The

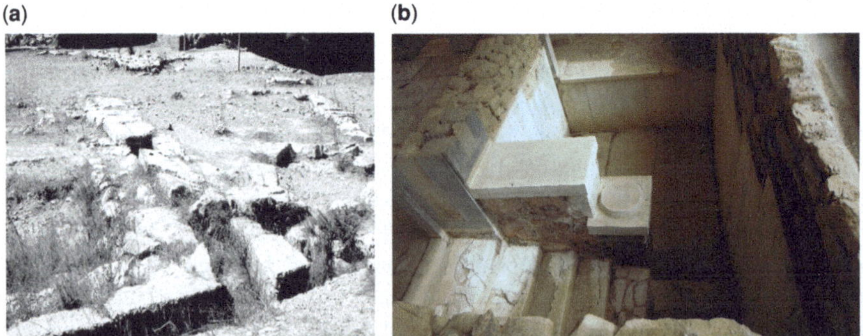

Figure 3.2 (a) Parts of sewerage and drainage systems in the Zakros palace and (b) Minoan bath in Phaistos palace.

(a)

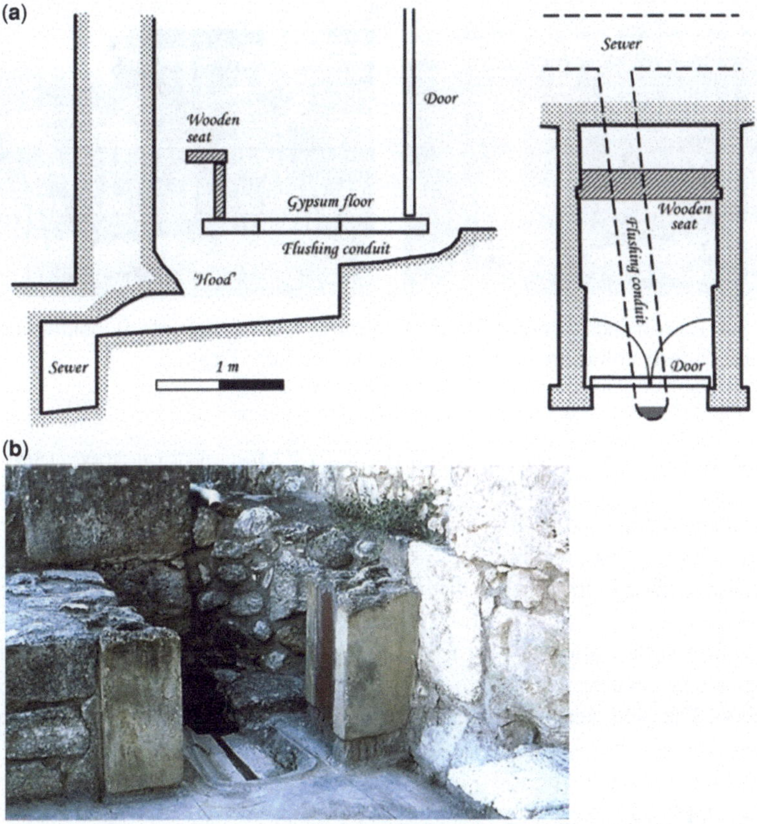

(b)

Figure 3.3 Section and plan of ground-floor toilet in the residential quarter of palace of Minos (a) (adapted from Antoniou *et al.*, 2016) and (b) General view. (http://www.nature.com/news/the-secret-history-of-ancient-toilets-1.19960).

existence today of several Minoan archaeological sites is mainly due to their very advanced drainage systems. In addition to sewers, bathrooms (Figure 3.2b), toilets and other sanitary structures were considered necessary in most Minoan palaces (Angelakis & Rose, 2014). One of the first known toilets worldwide is that in Knossos (Figure 3.3). A similar one was discovered in a house near the palace at Malia.

3.3 HISTORICAL TIMES

3.3.1 The classical and hellenistic periods (*ca.* 490–67 BC)

After *ca.* 1420 BC, the Minoan civilization began to decline and was overshadowed technologically by the development of the Mycenaean civilization in mainland Greece. There is a possibility for continuity between the Minoan and Mycenaean

Figure 3.4 Hellenistic cisterns; (a) at Eleutherna town (excavated) inside view and (b) central cistern at the ancient town of Lato in eastern Crete.

civilizations, through another civilization, that of Minyans (*ca.* 2000–1500 BC). According to (*ca.* 65 BC–AD 23), the Minyans had managed to drain the lake of Kopais in the 15th century BC, using hydro-technologies similar to those of the Minoans. Also, these technologies, despite the darker periods that followed, continued and developed further in the Archaic, Classical and Hellenistic periods (Angelakis *et al.*, 2012, pp. xii, 560; Angelakis & Rose, 2014). In Crete, during the Classical and Hellenistic periods, technologies related to water supply, sanitation and hygiene in general, were further developed based on Minoan knowledge and experience. The achievements of this period, regarding the design, the construction and operation of water resources projects, were impressive and more sophisticated.

During the Hellenistic period (*ca.* 323–67 BC), significant developments relevant to water supply and to a hygienic lifestyle were achieved in Crete, building on Minoan knowledge (Dialynas *et al.*, 2006). During this period, impressive accomplishments were achieved in hydraulics in Crete, such as the construction and operation of aqueducts, cisterns, wells, harbors, water supply systems, baths, toilets, and sewerage and drainage systems. Examples of significant hydraulic works exist in several Hellenistic towns, for example, Eleutherna, Lato, Dreros and Priansos. Two such examples of Hellenistic cisterns are shown in Figure 3.4.

3.3.2 The Roman period (ca. 67 BC–AD 330)

Later on, during the Roman period, Crete was involved in the Mithridatic wars, initially repelling an attack by Romans in 71 BC. Nevertheless, a ferocious three-year campaign soon followed under Quintus Caecilius Metellus, equipped with three legions. During the Roman occupation, Crete was incorporated with Cyrene (the African Libya) into a single Roman province with Gortys as the capital of the island. Public buildings, often with fine mosaics, toilets, sewers, drains, and other hydraulic works, were established in many of the main cities of the island including Gortys, Ierapytna, Aptera, Lyttos, and Lebena. In addition, public engineering works like buildings and roads and the first larger scale aqueducts and cisterns were constructed such as Gortys' aqueduct and cisterns in Aptera (Davaras, 1976).

The Romans did not add much to the Greek knowledge, although, the invention of concrete by Romans, called '*opus caementitium*', enabled the construction of longer canals, huge aqueducts, and long tunnels in soft rocks at lower costs (Fahlbusch, 2010). Thus, the Romans built 'mega' water supply systems, including many magnificent structures. The advanced water and wastewater technologies developed in Minoan and Hellenistic Crete were expanded and improved during the Roman domination of the Greek world. The achievements of this era, which met the hygienic and functional requirements of ancient cities, were so advanced that they could only be compared to the modern urban water systems which developed in Europe and North America in the second half of the 19th century (Mays *et al.*, 2007). Moreover, it should be noted that hydraulic technologies in ancient Greece were not limited to urban water and wastewater systems. The supply of urban water was even more progressive, as numerous aqueducts, cisterns, wells, and other water facilities indicate (De Feo *et al.*, 2010).

In Roman Crete there were several aqueducts such as in Axos, Chersonessos, Falassarna, Minoa, Kissamos, and Gortys (Figure 3.5a). A very impressive Roman aqueduct of a total of 22 km length has been described in ancient Lyttos (Angelakis *et al.*, 2012, pp. 227–258). Its water source was located on the west flank of Oropedio Nissimou mountain (summit, 1,148 m high), at a location called Kournias, a few kilometres south of the village Krassi, located at an altitude over 600 m.

In addition to aqueducts, several cisterns have been found in Dictynna, Lappa, Rhizenia, Elyro, Aptera, and Gortys (Figure 3.5b). The town of Aptera, located south of Souda bay, in Chania, is regarded as one of the most significant townships on the island during the Hellenistic and Roman periods.

During the Roman period, Crete participated in the Mithridatic wars, repelling the attack of the Romans (71 BC). However, following a three-year campaign under Quintus Caecilius Metellus, the island was occupied. During the Roman

Figure 3.5 (a) Remnants of Roman aqueduct in the ancient city Gortys and (b) Roman cistern (rectangular cross-section) at the Praetorium of Gortys.

occupation, Crete was a single Roman province, with Gortys as capital. In large cities, such as Gortys, Ierapytna, Aptera, Lyktos, and Lebna, in several public buildings mosaics of sophisticated technique and toilets with flushing mechanisms, water supply, irrigation and drainage networks and other hydraulic works were constructed. However, the Romans did not significantly further advance the basic knowledge on urbanism, building and especially hydraulics, developed earlier by the Minoans and the ancient Greeks. In contrast, the Romans significantly upgraded the scale of the hydraulic works, such as the Gortys and Hersonissos aqueducts and the Aptera reservoirs.

3.4 THE BYZANTINE PERIOD (*CA.* AD 330–1204)

For most of the period from AD 330 to 1204, Crete was part of the Byzantine Empire. 'Chandax' (present-day city of Iraklion) was the headquarters of the Duke of Crete. The technologies used for water supply of the cities were slightly improved during the Arabic period (ca. AD 827–961). There is evidence that the Fountana spring was used by the Romans and the water supply of Iraklion city was improved (Spanakis, 1981). However, the central part of the Fountana aqueduct was rebuilt during the Egyptian Period. At that time (1839), the tunnel in Skalani was reconstructed with a cross section of 1×2 m^2 and a length of 1,150 m. It appears to be one of the longer Roman tunnels in the Mediterranean region. The aqueduct in Agia Irini was built during the Egyptian period. The spring is located 5 km from Knossos, at an altitude of 220 m and the entire supply of good quality water is still used today to provide water for the city of Iraklion (Angelakis & Spyridakis, 1996). Thereafter, from the beginning of the Venetian period (AD 1204–1668), particular attention was paid to water issues. Water cisterns were the major water technologies developed in Crete during that period (Figure 3.6a). Also, several water wells have been discovered in rich homes in Iraklion (Figure 3.6b). At the end of the Byzantine period Crete fell into the hands of the Venetians.

(a) (b)

Figure 3.6 Byzantine hydro-structures: (a) Cistern (of rectangular cross-section) in the Areti monastery in eastern Crete used mainly for irrigation and (b) Mouth of water well in the Historical Museum of Iraklion.

3.5 RENAISSANCE AND MODERN TIMES

3.5.1 The Venetian period (ca. AD 1204–1669)

From the beginning of the Venetian period (*ca.* AD 1204–1668), the Duke of Crete was in charge of the maintenance of the water tank of the Duke's palace, explaining that: '*...because water is a high necessity for the palace and the family of the Duke of Crete*'. Venetians gave much attention to water supply issues (Strataridakis *et al.*, 2012). During this period, large-scale water projects were also realized (e.g., Morozini's aqueduct and cisterns in Venetian Rethymnon and Grambousa).

Francesco Morosini, *Provveditore Generale* (Superintendent) (1625) of the city of Candia (present-day Iraklion), decided to carry out an ambitious plan for an effective water system in the city. He planned to connect several water springs together into one big aqueduct. Morosini's idea for this plan was based on a report by Latino Orsini (*ca.* 1530 to *ca.* 1580), who had referred in detail to the water springs on the Youktas Mountain.

There were two points that led Morosini to the enactment of his idea: on one hand, there was the elevation difference between the Youktas and Iraklion and, on the other, the influence of good quality water springs. Approval was given for a huge expenditure of 13,000 *regals*, by the Council members of the Venetian Kingdom of Crete. In 1627, the project began and progressed at remarkable speed, considering its magnitude as well as the conditions of the times. The conduit was being constructed by a number of workers spreading out along a 15.5 km distance from the initial water spring to Candia Square (Strataridakis *et al.*, 2012). The study of the route and the slopes that the aqueduct could follow was a big project in itself. Thousands of workers (employees, together with villagers in *angaria*, a form of compulsory service) were utilized: builders, sculptors (for the fountains in the city), and other specialized technicians worked for the project (Figure 3.7a). Well-known engineers were also employed, such as, Zorzi Corner, Rafaele Monanni and Francesco Basilicata (Spanakis, 1981).

(a) (b)

Figure 3.7 (a) Remnants of the Venetian aqueduct (Morozini) in the area of Karidaki, Iraklion and (b) Central cistern of two-linked chambers in the Arsenal Nuovi in Iraklion.

The largest cistern in Iraklion during the Venetian period was that built adjoining the Arsenal Nuovi (near to Venetian port) of a capacity 450 m^3 (Figure 3.7b). The cistern consists of two oblong vaulted chambers linked by arched openings. Light and air enter the cistern through large light-shafts in the top of the vaults. In addition all the daily garden requirements of the port (about 7 m^3 of water/d) are covered from another underground cistern built by Venetians. Finally, several cisterns and fountains were instigated throughout the island during the Venetian period.

3.5.2 The Ottoman and the Egyptian periods (ca. AD 1669–1898)

In the Ottoman period, water was closely linked to religion; in every mosque there was a fountain for the religious needs of the Moslems (Spanakis, 1981). Fountains were built in most Cretan cities, many of which still exist, such as in Ierapytna (at the entrance of the Mosque), Chania (in the Archaeological Museum) and Iraklion (in the area of the Archaeological Museum (Angelakis, 2015). The Egyptian period (AD 1830–1840) was a continuation of Ottoman rule. During this period, the hydraulic constructions, built mainly by the Romans and the Venetians, were maintained and operated (Angelakis *et al.*, 2012).

During the first period of the Ottoman conquest, many Turkish neighborhoods in major cities had fountains (Angelakis, 2015). However, with Christians and Moslems living side by side, there was not enough water. Only a few houses had running water or cisterns and they usually belonged to Ottoman officials. Evligia Çelebi noted characteristically that during the Ottoman period in Iraklion there were 70 drinking fountains (Strataridakis *et al.*, 2012). One such fountain is shown in Figure 3.8a. Also a water bridge, part of the Fountana aqueduct which was reconstructed by the Egyptians is shown in Figure 3.8b.

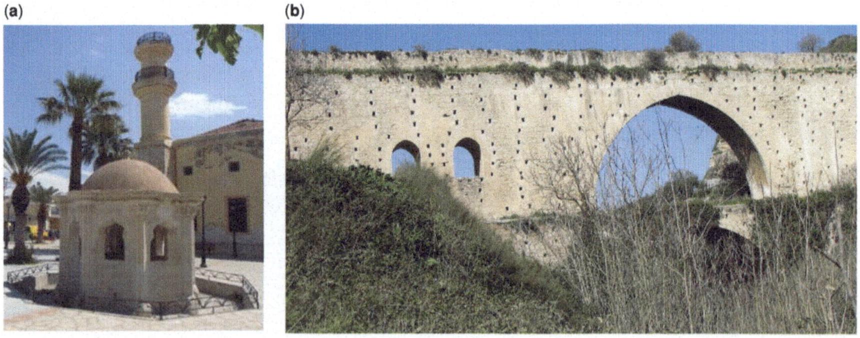

Figure 3.8 (a) Ottoman fountain outside of the Moshe in Ierapetra city and (b) Foundana aqueduct: remains of aqueduct at Aghia Irini, Iraklion.

3.5.3 Modern times

By the end of the 19th and beginning of the 20th century, the Cretan state was established and modern water technologies were initiated, as in other parts of the world. They were based on the technologies of the past as well as on deep wells, pumps, pipes, etc (Angelakis & Vavoula, 2012). This development continued to advance after Crete's merger with mainland Greece in 1913, and even more after the World War II and the following Civil War.

3.6 DISCUSSION AND CONCLUSIONS

The climatic and geohydrological conditions in the island are highly differentiated from north to south and east to west. Thus, water availability is spatially and temporally uneven and it is clear that water technologies, developed through the centuries, such as aqueducts, cisterns, wells, and others were created and adapted to the local conditions, presenting an ecological, sustainable system with considerable potential in water supply. Certainly, many of today's advances were not known in antiquity but, to some extent, differences are noted in the apparatus and scale of applications used rather than in the application of fundamental principles (Angelakis *et al.*, 2012).

Regarding the evolution of water science and engineering in Crete, it appears to be non-linear and often characterized by discontinuities and regressions. Cretan hydro-technological achievements were not totally forgotten during the Greek Dark Ages (ca. 1150–800 BC). 'Bridges' from the past to the future are always present, albeit oftentimes invisible to those who cross them (Angelakis *et al.*, 2012, pp. xii, 560). Thus, in addition to many ancient constructions that have been continuously or intermittently in operation up to modern times, substantial information from written ancient Greek sources has also been preserved, including such specifics as detailed contracts between the public and constructors of hydraulic works (Angelakis *et al.*, 2012). A characteristic example of survival is provided by the several types of cisterns developed in Minoan and Mycenaean times which are still widely used in many waterless Greek islands. A continuous evolution of the engineering of cisterns that bridges several historical periods is clearly indicated (Antoniou *et al.*, 2006).

Ancient Cretans lived in harmony with nature and their environment; those that did not, failed. Local water supply sources were first used by the Cretans of the Minoan Era. When these were exhausted, local and temporal transfers were instituted and the necessary hydraulic structures built. In this context it is noteworthy that the Minoans knew all the necessary basic principles and physical parameters. Those basic Minoan hydraulic works were further improved by Greeks during the Classical and Hellenistic periods. The Romans did not add much to this fundamental knowledge (Fahlbusch, 2010); however, the scale of application and of the apparatus used were highly effective during the Roman period. By comparison of the drainage and sewerage techniques developed during the historical times to the elaborate prehistoric systems in the Indus Valley and Knossos, dating to a couple of millennia BC, Wilson's conclusion (2000) is justified, that 'the individual elements of drainage systems appear relatively early and remain largely unmodified until the Middle Ages'.

In conclusion, the diachronic development of water science and technology in Crete is characterized by discontinuities and regressions. Technological achievements, such as those during the Minoan era or the Classical and Hellenistic times, were not abandoned despite the various occupants of the island. There were bridges that linked the past with the present, which often were not visible. Some of the most prominent features of the Minoan civilization were the hydraulic and architectural constructions and operation of the water and sewerage systems of waste- and rainwater in the palaces and the cities. In most Minoan palaces and cities, nothing is more remarkable than the still existing complex and functional hydraulic systems (Angelakis & Spyridakis, 2013). It can be concluded that the ancient hydrologists and engineers, especially the Minoans, were aware to some extent of the basic principles of the sciences of water resources and the environment, allowing us to propose that:

(a) No fundamental differences exist in the design and construction of water works from the Minoan era until today. The key differences are the scale, the equipment available and the experience and knowledge gained.
(b) Selected technologies in a given time and place, varied not only according to time periods but were also depending on the hydrological conditions in each region and of course the availability of water resources.
(c) Ancient hydro-technologies were remarkably designed and adapted to the environment.
(d) Their working time, in many cases, reaches millennia.
(e) The forecast for future increased water needs was also evident.
(f) Integrated water resource management, based on balancing availability and demand was well-thought-out.
(g) Finally given the above, the current concept of sustainability as a planning principle for water resource projects may need to be revised. That is, in the definition of sustainability, as a production model which aims at the best economic result for both the human beings and the present and future natural environment.

ACKNOWLEDGEMENT

Part of this work was presented in the Conference on Protection and Restoration of the Environment XI, held on 3–6 July, 2012 in Thessaloniki, Greece.

REFERENCES

Alexiou S. (1964). Minoan Civilization. S. Alexiou and Sons, Iraklion, Greece (in Greek).

Angelakis A. N. (2015). The history of fountains and relevant structures in Crete, Hellas. *International Journal of Global Environmental Issues*, **14**(3/4), 200–215, http://dx.doi.org/10.1504/IJGENVI.2015.071865

Angelakis A. N. (2016). Evolution of rainwater harvesting and use in Crete, Hellas through the millennia. *Water Science and Technology: Water Supply*, **16**(6), 1624–1638, doi: 10.2166/ws.2016.084

Angelakis A. N. (2017a). Hydro-technologies in Minoan Era. *Water Science and Technology: Water Supply*, **17**(4), 1106–1120, doi: 10.2166/ws.2017.006

Angelakis A. N. (2017b). Urban waste- and storm-water management in Greece: past, present and future. *Water Science and Technology: Water Supply*, **17**(5), 1386–1399, doi: 10.2166/ws.2017.042

Angelakis A. N. and Rose J. (eds.). (2014). Evolution of Sanitation and Wastewater Management through the Centuries. IWA Publishing, London, UK, p. 529.

Angelakis A. N. and Spyridakis S. V. (1996). The status of water resources in Minoan times – a preliminary study. In: Diachronic Climatic Impacts on Water Resources with Emphasis on Mediterranean Region, A. Angelakis and A. Issar (eds.), Springer-Verlag, Heidelberg, Germany, pp. 161–191.

Angelakis A. N. and Spyridakis D. S. (2010). Water supply and wastewater management aspects in ancient Hellas. *Water Science and Technology: Water Supply*, **10**(4), 618–628.

Angelakis A. N. and Spyridakis S. V. (2013). Major urban water and wastewater systems in Minoan Crete, Hellas. *Water Science and Technology: Water Supply*, **13**(3), 564–573.

Angelakis A. N. and Vavoula G. (2012). Evolution of urban hydro-technologies in Crete, Hellas through the centuries. Protection and Restoration of the Environment XI, CD Proceedings and Book of Abstracts, 3–6 July 2012, Thessaloniki, Greece, p. 323.

Angelakis A. N., Koutsoyiannis D. and Tchobanoglous G. (2005). Urban wastewater and stormwater technologies in ancient Greece. *Water Research*, **39**, 210–220.

Angelakis A. N., Dialynas M. G. and Despotakis V. (2012). Evolution of water supply technologies in Crete, Hellas Through the Centuries. In: Evolution of Water Supply Throughout Millennia, A. Angelakis et al. (eds). IWA Publishing, London, UK, pp. 227–258.

Antoniou G., Xarchakou R. and Angelakis A. N. (2006). Water cistern systems in Greece from Minoan to Hellenistic Period. Proceedings of 1st IWA International Symposium on Water and Wastewater Technologies in Ancient Civilizations, 28–30 October 2006, Iraklion, Greece, pp. 457–462.

Antoniou G., Kathijotes N., Angelakis A. N. and Spyridakis D. S. (2014). Historical development of technologies on water resources management and rainwater harvesting in the Hellenic Civilizations. *International Journal of Water Resources Development*, **14**(8), 1–14.

Antoniou G., De Feo G., Fardin F., Tamburrino Tavantzis A., Khan S., Fie T., Ieva Reklaityte I., Kanetaki E., Zheng X. Y., Mays L. and Angelakis A. N. (2016). Evolution of toilets worldwide through the millennia. *Sustainability*, **8**(8), 779; doi: 10.3390/su8080779

Broodbank C. and Strasser T. (1991). Migrant farmers and the Neolithic colonisation of Crete. *Antiquity*, **65**(247), 233–245, doi: 10.1017/s0003598×00079680

Davaras C. (1976). Guide to Cretan Antiquities. Noyes Press, Park Ridge, NJ, USA.

De Feo G., Laureano P., Drusiani R. and Angelakis A.N. (2010). Water and Wastewater Management Technologies through the Centuries. *Water Science and Technology, Water Supply*, **10**(3), 337–349.

Dialynas M. G., Lyritztis A. G. and Angelakis A. N. (2006). Historical Development of Water Supply of the City of Iraklio, Greece. Proceedings of 1st IWA International Symposium on Water and Wastewater Technologies in Ancient Civilizations, 28–30 October 2006, Iraklion, Greece, pp. 671–676.

Fahlbusch H. (2010). Introduction: political situation, attitude to the water and necessity of transfers. In: Water Engineering and Management in Classic Civilizations. Water Engineering and Management through Time–Learning from History, E. Cabrera and F. Arregui (eds.), CRC Press, Taylor and Francis Group, New York, Leiden, London, UK.

King K. R., Ozcan S. S., Carter T., Kalfoglu E., Atasoy S., Triantaphyllidis A., Kouvatsi C., Lin A. A., Chow C-E. T., Zhivotovsky L. A., Michalodimitrakis M. and Underhill P. A. (2008). Differential Y-chromosome Anatolian influences on the Greek and cretan neolithic. *Annals of Human Genetics*, **72**, 2015–214, doi: 10.1111/j.1469-1809.2007.00414.x

Koutsoyiannis D., Zarkadoulas N., Angelakis A. N. and Tchobanoglous G. (2008). Urban water management in ancient Greece: legacies and lessons. *Journal of Water Resources Planning and Management*, **134**, 45–54.

Markonis Y., Angelakis A. N., Christy J. and Koutsoyiannis D. (2016). Climatic variability and the evolution of water technologies in Crete, Hellas. *Water History*, **8**, 137–157, doi: 10.1007/s12685-016-0159-9

Mays L. W. (ed.). (2010). Ancient Water Technology. Springer, Dordrecht Heidelberg, London, UK and New York, USA.

Mays L. W., Koutsoyiannis D. and Angelakis A. N. (2007). A brief history of urban water supply in antiquity. *Water Science and Technology: Water Supply*, **7**(1), 1–12.

Spanakis S. (1981). The Water Supply of Iraklion, 828–1939. The Technical Chamber of Greece, Iraklion, Greece (in Greek).

Strataridakis A. I., Chalkiadakis E. G. and Gigourtakis N. M. (2012). The historical Development of Water Supply to Iraklion, Crete, Greece. In: Evolution of Water Supply Throughout Millennia, A. Angelakis *et al.* (eds.). IWA Publishing, London, UK, Ch. 18, 467–496.

Tsonis A., Swanson A., Sugihara K. L. and Tsonis P. A. (2010). Climate change and the demise of Minoan civilization. *Climate of the Past Discussions*, **6**, 801–815.

Wilson A. I. (2000). Drainage and sanitation. In: Handbook of Ancient Water Technology, Ö. Wikander (ed.), E. J. Brill, Leiden, The Netherlands, pp. 151–179.

Chapter 4

Region, nation and culture on the British waterways, 1761–1894

Jodie Matthews

ABSTRACT

This chapter considers the ways in which writing about Britain's canals, from the Duke of Bridgewater's Canal to the opening of the Manchester Ship Canal, took influence from, framed, and contributed to contemporaneous discourses of region and nation. It proposes that, while schemes and management were often local affairs concerned with specific regional trade and industry interests, successes were considered to be *British*, with engineering and organisational innovation seen as national points of pride. Writing about the waterways contributed to the overwhelming sense in this period that Britain was a modern (perhaps *the* modern) nation, despite its ancient institutions and the fact that it often followed the lead of others in canal technology. The chapter thus takes the well-understood early history of the waterways and considers the ways in which they were represented in relation to discourses of region, nation, and modernity of the eighteenth and nineteenth centuries, using interpretive methodologies from historical, literary, and social/cultural studies to interrogate a national social hydrological discourse. It connects specialist waterways history to the larger question of nationhood to demonstrate the ways in which an understanding of historical representation offers insight into the cultural dynamics of water management.

Keywords: British waterways, Bridgewater Canal, Manchester Ship Canal, Region and Nation, early history

Editor: Xiao Yun Zheng
doi: 10.2166/9781789062045_0051

4.1 INTRODUCTION

In the year 1797, the *True Briton* magazine published an article almost breathless with excitement about the potential for a 'Great Commercial Canal', establishing inland navigation using existing river boats 'between the Ports of London, Hull, Liverpool, and Chester'. Such a scheme, the article explained, would open up new markets for 'the increasing trade of the Potteries' and connect regions 'abounding with minerals' such as coal, lime, copper, and lead, to manufacturing districts. The scheme was, the *True Briton* emphasised, 'an object of great *National* importance!' Further, supporting the scheme was described as the duty of 'every man who has the real interests of his Country at heart' ('The Great Commercial Canal', *True Briton*, 23 November 1795). The fact that the article appeared verbatim in *The Times* a month later indicates that this was more the work of the canal promoters than authentic journalistic enthusiasm, but their use of nationalist rhetoric suggests that this was a convincing strategy for garnering financial and legal support for the canal ('The Great Commercial Canal', *The Times*, 24 December 1795). An eighteenth-century pamphlet on 'the great utility of a circular and other inland canal navigation in Britain' suggested that such a scheme offered advantages 'both to State and Subject' (Hibbard, 1796). Looking to such patriotic calls, this chapter examines writing about British canals and waterways from the period in which canal development began in earnest in this country up to the opening of the Manchester Ship Canal in order to consider the ways in which the waterways network was represented in relation to discourses of region, nation, and modernity of the eighteenth and nineteenth centuries. I refer to the period 1761 to 1894 as an 'extended canal age', a useful slice of British history for interrogating the interrelationships of waterways history with other cultural discourses.

The introduction proceeds by highlighting some aspects of the development of a national waterway network in Britain, and the scholarship and opinion that have shone a light on regional and national concerns. It then considers Jamie Linton's ideas about people's 'cultural' connections water in this context, an approach that takes account of the fact that 'the history of water is a story of how people have drawn meanings, ideas, representations, and powers from water' (Linton, 2010). The introduction also clarifies what is understood by 'nation' in this period in Britain. The chapter moves on, with this understanding, to discuss region and nation in relation to four specific canal projects: the Duke of Bridgewater's Canal; the Grand Junction Canal (developed with similar motivations to those described by the *True Briton*, above); one Scottish example, the Caledonian Canal; and, finally, the Manchester Ship Canal. Rather than a survey of all representations on the subject, the texts selected within each case study offer an indication of the ways in which the discourses of region and nation are mobilised when describing canals. In conclusion, I assert that insight into the cultural dynamics of water management can be generated by locating waterways history within broader discursive sweeps, such as nationhood, but also that an understanding of nationhood is nuanced by investigating the representation of the waterways.

In 1848, George Beadon, commander of a venerable defender of the nation, the Royal Navy, noted his astonishment that it took his 'enterprising countrymen' until 1761 to fully exploit the possibility of canals, given early Chinese and Egyptian adoption of this technology (Beadon, 1848). However, as Philip Bagwell explains, 'the improvement of Britain's inland waterways was a continuous process dating back at least as early as the second half of the sixteenth century', and the chapters of John Blair's edited collection point to much earlier improved waterways and canals in England for transport and other purposes from the eleventh century (Bagwell, 1988; Blair, 2008). The later 'improvement' constituted widening and deepening rivers such as the Aire and Calder and the Medway, vital to the development of local industry and facilitating the exploitation of coalfields. Improved rivers were still unpredictable, though, and were an inefficient means of carrying large volumes of fuel and materials.

The burden of initiating canal development usually fell, as Bagwell points out, to the local landowners and entrepreneurs who had most to gain. It was they who called promotional meetings, had pamphlets published, employed surveyors, and injected capital. The Duke of Bridgewater, whose canal forms the first case study below, 'was merely the most famous of a long line of distinguished local sponsors of canals' (Bagwell, 1988). This is not to say that canals were exclusively local concerns: 'in the case of the Ellesmere canal project, launched in 1791–92, investors came from as far afield as Derby, Leicester, Northampton, Coventry and Birmingham' (Bagwell, 1988). This is not to say, either, that local support would ever be enough for a new canal: before a single spade could cut the earth, a Canal Act would need to pass through the national Parliament. Anthony Burton has shown how the narrative of canal development offers intriguing insight into the institution of Parliament before the Reform Act, and how difficult it was for canal promoters in manufacturing areas, which often had no Members of Parliament, to find support for their proposal in Westminster. The sheer number of canal bills being presented to Parliament drove it in 1792 to design a new process whereby proposals were advertised in the *London Gazette* and local papers, maps deposited with the local Clerk of the Peace, and full details of costs, the route, and assenting and dissenting landlords listed (Burton, 2005).

Localised, piecemeal promotion of canals delayed and complicated the development of a truly national network. There were numerous canal companies, each with their own boards of directors, management, and charges; nothing was centralised. There were also physical constraints to understanding the waterways as a national system: there was no standard broad gauge for canals, so 'carriers were precluded from exploiting the undoubted advantages of the long haul, by water transport, of large consignments of goods' (Bagwell, 1988). Nevertheless, the canals, according to Beadon with typical Victorian bombast, 'developed the resources of the nation, facilitated commerce, spread civilisation, and helped to raise Old England to its present eminence among the nations of the earth' (Beadon, 1848). Bagwell agrees, at least, that 'the canals contributed handsomely to [...] the levelling out of regional differences in prices and the cheaper distribution of the food of the nation' (Bagwell, 1988). In 2015, Liz McIvor took up this reading of the canals as a national resource with a book accompanying a BBC television series titled *Canals: The Making of a Nation*. She asserts that, despite the difficulties, canals 'developed a network between regions, a communication between towns and cities that didn't exist before' (McIvor, 2015).

Throughout all of this development, as Linton predicts, 'people have drawn meanings, ideas, representations, and powers from water'. He asserts that we cannot ignore 'ignore water's *cultural* dimensions, as in the myriad ways that water articulates with people to produce different meanings and different kinds of relationships' (Linton, 2010). Similarly, Purdue *et al.* note that 'canals appear as a structural element of the socio-political organisation of communities and their territory on various temporal and spatial scales' (2015). The case studies that follow are a brief introduction to the ways in which the waterways produce a sense of the nation. They 'materialize hydrosocial relations', as Linton would have it, because of the power and authority invested in building, using, and maintaining them. After all, the 'uses to which water is put' and 'the distribution of benefits that accrue from its allocation are central to any social order', an order that is 'maintained through discourses, laws, and tacit as well as formal rules that reinforce certain ideas and meanings of water in any given society' (Linton, 2010). In other words, we can understand the discourses that construct an idea of Britain's waterways as implicated not just in the politics of who benefits from industrialisation, and the politics of who and what constitute the industrialised nation, but also of how water itself is conceived in a nation cut by canals. After Linton, then, the approach that these case studies take is to see the representations of the canals in question as flowing into *and* out of a national social hydrological discourse. But how was Britain as a nation understood more broadly in the period under consideration here, roughly 1761 to 1894?

This chronological focus corresponds with the Industrial Revolution and the industrialised existence that followed, and with war, imperial ambition, and dominance. Any summary of nationhood during this particular span of time will necessarily be a broad-brushstroke history. Indeed, any history of the nation is also a comment on the present, and it should be noted that the concept of the British nation is under simultaneous strain and celebration in the twenty-first century. It almost goes without saying that the men and women who built, lived on or near and laboured on the canals will have had a very different conception of the nation than the aristocracy whose land the canals bisected and the commentators whose writing comes under scrutiny here. As Douglas Hay and Nicholas Rogers point out in a survey of eighteenth-century English society, 'those who wield power, in most societies, [...] are indifferent to much of what goes on in the lives of their fellow citizens' (1997). When Hay and Rogers consider those citizens, they think about the 'material legacies of their labour', including the canals they built. For these authors, then, understanding England of the 1700s is a process that takes in extant canals as well as the ideas that survive from that period, such as its politics and concept of nation. Particularity and local practice (sustained by the poor laws that tied individuals to the parish in which they were born) were 'immensely important in the social life of the eighteenth century', but transport, political discourse, and 'national consciousness' eroded much of this particularity as that century ended. Nevertheless, 'the phrase "the country", in local usage, generally meant the shire rather than the nation' (Hay & Rogers, 1997). Developments in transport – canals and turnpike roads – did contribute to a sense that the nation was now a space one could traverse and know. Even before these improvements, the population was internally mobile for purposes of trade and work, but generally did not travel very far. The rise of a provincial press recirculating metropolitan news (and, increasingly, a population who could read it) contributed to knowledge of the nation. Water shaped the 'mental landscape of the nation' in other profound ways: 'the American colonies, the Caribbean sugar islands, Hudson Bay and Newfoundland, West African slaving posts, and Indian and Chinese ports' meant that 'Greater Britain' incorporated overseas territories, and also brought thousands of Britons into contact with people of other cultures, religions, and ethnicities against whom they defined themselves (Hay & Rogers, 1997). Such encounters did not just happen overseas; Liverpool, for instance, was a detention centre for foreign prisoners of war from America and Europe; individuals and institutions were in close contact with them (Hill, 2017). Diana Ascott and Fiona Lewis found eighteenth-century Liverpolitans born in Antigua and Virginia in parish registers (2000). The port recruited workers from the Americas, Europe, and Scandinavia. German merchants (mainly from northern Germany, in particular Bremen) began to settle in Liverpool from the mid-eighteenth century (Lee, 2012). John Belchem has detailed the Irish migrant experience in Liverpool (for instance, see Belchem, 1999, 2005). Britain was, then, a place with nationally and ethnically diverse port city populations.

Closer to 'home' and internal colonialism, lands in Wales and Ireland had been colonised by the Anglo-Normans since the twelfth century, and Wales had been politically subordinated to England since the sixteenth. Scotland and England were joined in the Union in 1707. The Act of Union also, notably for the subject of this chapter, did away with all internal customs duties and trade barriers. Given the focus on such barriers today, as Britain painfully extricates itself from the European Union, the connections between identity, nation, and customs duties can be understood as surprisingly strong and definitive. After the defeat of Scottish aspirant to the throne Charles Edward Stuart and his supporters – most bloodily at Culloden in 1746 – Parliament deliberately eradicated forms of Scottishness that did not cohere to an English vision of the nation. At the end of the eighteenth century, about one-fifth of the population of the British Isles did not speak English as their first language, with many of those speaking no English at all (Hay & Rogers, 1997). All of this detail is by way of suggestion that the nation as a singular idea was a contingent construction (a concept long established; see, for instance, Anderson, 2006; Bhabha, 1994) strengthened by its deployment in discussion of the waterways and other causes calling in and towards a national 'voice' (Morecroft, 2017).

The preface to Linda Colley's landmark study of the British nation from the Act of Union to Victoria's coronation reminds us that 'Britishness was a surprisingly rapid and always partial invention' which never obliterated other and older identities (1994). She is clear in her conclusion that the 'coming together' of the nation was thanks, in large part, to economic conditions of the eighteenth and early nineteenth centuries, a compact island held together by canals and turnpike roads, enabling the flourishing of free trade. These economic conditions coincided with a politically Protestant identity at a time of war with a Catholic state (France). Colley assiduously avoids texts written by politicians and intellectuals which 'make the national sentiment of the past appear more monolithic and systematic than it ever was', while this chapter actively seeks out those representations which labour to shore up the discourse of Britishness through their celebration of the waterways in the four case studies that follow. It does not assert, as my nod to the working-class people of the waterways indicates, that British nationalism was all-pervasive; those with power and influence were those shouting loudest about the canals as a national endeavour. In its analysis of representation, the chapter necessarily examines archival material, and where a digitised version is available I have indicated in the reference list where readers might find it, even when I have, in many cases, consulted an original.

4.2 THE BRIDGEWATER CANAL

On 17 July 1761, the first ten-mile stretch of the Duke of Bridgewater's canal, designed to link the coalfields on his estate to the growing markets in Manchester and Liverpool, was opened. It is often understood (erroneously) as Britain's first modern canal and, for Bagwell, 'the opening of this canal was as dramatic a turning point in the history of British transport as was the [later] opening of the Liverpool and Manchester Railway'. It demonstrated both engineering skill and the economic potential of canals (Bagwell, 1988). Famously, the price of coal in Manchester immediately halved. Francis Egerton, the third Duke of Bridgewater, had visited the French Canal du Midi on his Grand Tour of Europe and employed James Brindley as the engineer to link Bridgewater's mineral wealth via water to the rapidly developing urban crucibles of the north. The agriculturalist Arthur Young, in his northern English travelogue published nine years after the first stretch of the canal opened, described this plan as ranking Bridgewater, a 'spirited young nobleman', among 'the most useful genius's [sic] of this or any age' (Young, 1770). The canal was completed through to Stretford and

Castlefield Wharf in Manchester in 1765, and work began to forge a link with the port of Liverpool. The magnitude of this engineering feat, and many disputes about land ownership, led to the Duke being (by today's value) approximately £2million in debt for a time ('A proud heritage'). The canal did turn a profit eventually, and his wealth at death in 1803 was around £600,000, with additional bequests (Fairclough, 2009). The fifth extension of the canal was secured in 1795, with a link to the Wigan branch of the Leeds and Liverpool Canal, integrating the Bridgewater with the watery web it had initiated.

The Telegraph newspaper pronounced the Duke, in 1795 'one of the most magnificently useful of all our fellow-citizens'. The Duke's wealth was applauded, in keeping with the politics of this publication, because of the 'infinite accommodation to the public' that his investment in the canal enterprise offered ('In and Out of Place', *Telegraph*, 7 October 1795). In other words, the canals enabled wealthy patrons to prove their worth as citizens because of the national good the waterways brought. Michael McCahill has examined the interplay of peers, manufacturers, and state during the industrial revolution, with Egerton a key example. Throughout the eighteenth century, the members of the nobility managing their own mineral properties dwindled, replaced by partnerships and leasing agreements, with a rare few 'adopt[ing] the latest techniques in exploiting this wealth' (McCahill, 1976). McCahill points out that it was 'their political resources [that] enabled lords to dictate the contents of canal legislation', moulding bills to suit their needs (1976). Another example of a noble early canal supporter was William, second Earl of Dartmouth, who 'secured the passage of the Birmingham Canal Bill for its committee' and felt a strong obligation to assist the Midlands manufacturers. Canals were, of course, beneficial to both the owners of mineral resources and those who used the minerals in manufacture, but peers might also gain political capital from an enriched local community (McCahill, 1976).

Beadon, writing about Egerton half a century later (and quoted above), noted that 'for nearly a century this great benefactor's example of firmness, and the ingenuity of Brindley [...] continues in lucrative operation'. He lamented that among '*Young England*' of the mid-nineteenth century, such an engineering marvel went unnoticed 'amidst railway topics'. For Beadon, it was a fault of the national zeitgeist that this national asset was, by then, underappreciated. In 1770, though, when agriculturalist Arthur Young was writing his travelogue, this particular navigation still induced wonderment, and Young notes 'the number of foreigners' who have come to have a look at it, the canal commanding 'the attention and admiration of all *Europe*' (Young, 1770). Indeed, the design and execution of the canal 'can never be sufficiently admired' because it 'improve[s] and adorn[s the] country' (Young, 1770). On the other hand, Robert Fulton remarked some three decades after the canal's completion that 'tunnels, aqueducts, reservoirs, and embankments, familiar to foreign nations, struck the astonished Englishman with wonder', a comment that suggests that patriotic assertions of Britain being at the vanguard of waterways technology were more rhetorical than factual, and setting Britons up as technologically parochial (Fulton, 1796). Even at this juncture, Fulton identified the national advantages of improving communications between districts, improvements afforded by the sort of canal navigation ushered in to England by the Bridgewater Canal; not only would agriculture and commerce improve, but 'happiness spread, in proportion as the facility of conveyance increases'. Each area 'transmits its improvements to the other, and each feels the beneficial effects resulting from the union', what he called 'a common flock'. He makes a case for such communications between the different districts of a nation rendering it 'more independent of its neighbours', making better use of its 'internal resources' (Fulton, 1796). England is not his model case, however: he points to Egypt, India, and China as leaders in canal communication.

For some authors, the opening of the Bridgewater Canal signalled British pre-eminence and could be held up as a source of national pride. For others, a sophisticated waterways network could strengthen any nation, and Britain followed the path of others in this respect. The effect that this canal had on the industrialisation of the region is still recognised culturally today: in a 'Great Exhibition of the North' held in summer 2018 to celebrate the North of England's 'pioneering spirit', Tyne & Wear Archives & Museums selected 100 objects to tell the history of the north ('Great Exhibition of the North', 2018). Included in this selection was a 'Starvationer' boat with exposed ribbed sides, used in the underground canals to bring coal from the coal face out onto the canal running between Worsley and Salford. The design is considered the forerunner of the narrow boat still widely in use. The appeal of this object, as an icon of northern history, points to the close relationship between the canal and a sense of regional identity. Indeed, significant funding (almost £1million) was awarded in 2017 to the organisation IVE to help young people aged 14–19 'gain a new relationship with and understanding of Yorkshire's historic mills, workshops, towpaths, tunnels, canals and locks' (Green, 2017). In the twenty-first century, there is enthusiasm within regional institutions and organisations for re-engaging public audiences with the cultural histories of water management. The links between the waterways and regional and national identities are not unique to this pioneering canal, however; similar ideas can be found in descriptions of a developing network, as the following example of the Grand Junction Canal demonstrates.

4.3 GRAND JUNCTION CANAL

The Oxford Canal, completed in 1790, provided a link between the growing canal network and London. However, the route was long and difficult to navigate in stretches. A new canal was conceived that would offer a more direct and reliable link between the capital and the Midlands and the north of England: the Grand Junction Canal. George Nugent-Temple Grenville, 1st Marquis of Buckingham, commissioned the initial survey for the route, became a major shareholder in the Grand Junction Canal Company, and was a principal sponsor of the 1793 Grand

Junction Canal Act (Petticrew & Austin, 2012). That the story of eighteenth-century canals is populated by the aristocracy, the new middle and professional classes (in this case, the banker William Praed and engineers William Jessop and James Barnes), and the working people who built and operated them shows just how integral the waterways are to a national history of British people of all social strata.

In 1819, John Hassell published his *Tour of the Grand Junction*. The text is an important one for the history of representation of the waterways because of its self-conscious relationship to the picturesque, an extremely influential cultural movement in Britain. Celebrated picturesque landscapes conformed to rules of composition, balance, and interest. Tourists sought out the sort of landscapes described by the likes of the Reverend William Gilpin, who from the 1780s published picturesque guides to areas of Britain. Contemporary commentators suggested that 'there was a patriotic dimension to seeking out the picturesque beauties of the British Isles', assisting 'the unification and modernisation of the different regions and nations of newly formed "Great Britain"' and 'the extensive circulation of both tourists and prints helped generate a new national consciousness, heightening the perception of the British Isles as an interlinked whole' (Thompson, n.d.). Hassell made the most of an emergent public appetite for artistic representations of industrial scenes, producing an 'industrial picturesque' (Constantine, n.d.).

Hassell finds the Grand Junction Canal 'directing its course through scenes of undiminished beauty, and replete with delightful prospects, uniformly picturesque, and sometimes grand'. Remarking on the building of canals, Hassell asserts that genius requires the 'fostering hand' of patronage to make it 'subservient to *national* advantages'. Inland navigation, he adds, is, to a manufacturing country, 'the very heart's blood and soul of commerce'. He praises those members of the aristocracy who have 'allow[ed] this great national undertaking to pass through their respective parks'. He suggests that 'it must stand as a monumental record, and example, of the urbanity and *amor patriæ*, these distinguished noblemen exhibited for the weal of their country' (1819; emphasis added). Hassell allies aristocratic civility in enabling canal development with patriotism and the greater national good. In praising forward-thinking modernity, he reaches for the Latin '*amor patriæ*', 'urbanity' (which is a borrowing from French), and 'weal', a word meaning wealth that comes from the Old English 'wela' (OED). It is as if, in Hassell's rendering, the landed gentry's benevolent attitude to this technological advance is one that can be traced throughout England's cultural history. 'Weal' has been used since at least the fifteenth century to mean the welfare of a country or community. Hassell's linguistic choices are designed to bolster the old British class system in a modern age – particularly relevant at the close of the Napoleonic Wars. The implication is that revolution, and the removal of the aristocracy, in Britain would undermine its manufacturing might and the benefits that this (supposedly) brings to all.

To consolidate this image of modern national harmony, Hassell's engravings present winding canals that might as well be rivers from their depiction, and those who laboured on the canals simply as picturesque figures in the landscape. His descriptions, too, conform to the ideological project of integrating the canal and an industrial picturesque into the appreciation of the national landscape. For instance: 'After passing these locks at Stoke Hammond, the navigation inclines to the north-east, and is buried between woods which line the base of some high hills, forming a perfect Sylvan appearance' (1819). The woods look like woods are *supposed* to look; all is well with the nation even as it changes radically from the forces of industrialisation. Of each place through which the canal passes, Hassell informs his readers of its Roman remains, ancient churches, or long-standing trades and produce. Such unchanging locales sit, in Hassell's representation, compatibly alongside manufacture, such as 'the potteries of Staffordshire (a grand national monument of British industry)'. Both the form and content of Hassell's *Tour* use the Grand Junction Canal as an instance of the nation using its ancient institutions and relations to move into a new, profitable era.

Better known for its touristic landscapes than the Grand Junction Canal is the only Scottish case study in this chapter, the Caledonian Canal, conceived and surveyed in 1773 and finally opened in 1822.

4.4 CALEDONIAN CANAL

The author of *A Trip through the Caledonian Canal* was hoping to admire the 'romantic' Scottish scenery with friends and relations aboard the cutter, *Whipper*, accompanied by another vessel, the *Snapper* (1861). (These two names are a joke: a 'whippersnapper' is a young, inexperienced, and, perhaps, presumptuous person). The work was not intended for a wide audience, and this sort of private joke is indicative of the expectation of a small and knowing readership; it was privately printed and authored by 'Bumps' – identified by the blogger Mick Vedmore as B. P. Stockman, personal assistant to the civil engineer George Robert Stephenson, who was himself a keen yachtsman and who seems to be the 'Commodore' of this tale (2017). The trip happened at the behest and thanks to the funds of the 'Commodore', and it may be that he wished to partly repeat that embarked on but uncompleted by his famous cousin, Robert Stephenson, shortly before his death in 1859, and from whom he inherited a considerable sum (Smiles, 1868). This voyage as described by 'Bumps' is not in any way intended to bind rhetorically the nation together, as previously cited works did, but to present it as a space for adventurous leisure, particularly for those who had done so much to literally engineer national industrial success. Indeed, one character in *A Trip through the Caledonian Canal* asks what the time difference is between Scotland and London, indicating the psychological as well as geographical distance between the two; almost all Scottish towns and cities had, in fact, adopted Greenwich Mean Time some years earlier. There is no ideological investment here in suggesting that Britain is a more coherent whole thanks to projects like the Caledonian Canal or the railways pioneered by the Stephensons. Half a century of industrialisation produces for Stockman a different, and much less picturesque, image of Britain to that painted by Hassell. Glasgow, for instance, 'although a magnificent city' is said to possess 'no particular attraction to the tourist. It is all commerce and manufacture.' Indeed, the tourist 'will leave Glasgow as speedily as possible, perhaps a little disgusted'. En route to Glasgow, the party stop at Newcastle, where one member of the group refers to the local dialect as 'infernal gibberish' ([Stockman], 1861). This is no doubt another joke for a closed audience, as this was where Stephenson was born. The discursive effect, however, is to play the distinctiveness of regions and the nations of the United Kingdom for laughs, rather than seeing them as a noble tapestry stitched together by road and canal as other texts described in this chapter have done.

Two hundred pages into the travelogue, the party finally reaches the Caledonian Canal, 'awe-stricken' by the scenery, and with good reason: the canal is more aptly described as a series of canals connecting four naturally navigable freshwater lochs. They are, however, underwhelmed by the canal itself. Expecting to feel 'reverence' for 'colossal granite, hewn and shaped in almost every form that the sculptor's art could produce', they instead consider the canal 'awfully dingy and dull in appearance' (1861). Earlier commentators were less wry and more earnestly appreciative, however, and, like Hassell, connected the usefulness of the canal to the patriotism of aristocratic patrons. The *Inverness Courier* noted, in December 1827, that the dues on the canal had been so high as to prevent shipping from using this 'safe and short cut through the Great Glen of Scotland' until Lord Colchester's visit to the Highlands resulted in them being lowered. The newspaper ascribes the 'change of system to the patriotic and intelligent nobleman' ('Caledonian Canal', 1827). Despite the intervention of Lord Colchester in this instance, Urquhart Forbes noted in 1908 that the Caledonian was 'notable as the only British waterway which has been constructed entirely at the cost of and has always remained under the control of the state' (Forbes, 1908). Like many canal projects attempting to gain favour for their development, the Caledonian was described as a 'great national undertaking' and 'of the greatest importance to the nation generally' ('Caledonian Canal', 1840). The canal was not, as an 1824 article in the *Caledonian Mercury* pointed out, only useful to trade within Britain. It cited an instance of 'a vessel freighted at Riga [Latvia] with flax seed passing through direct to Londonderry [Ireland]. The same article cites the usual time to make the passage along the canal as three to four days; a stretch of water linking two seas also brings east and west coasts of Britain practically closer to each other ('Caledonian Canal', 1824).

One of the initial motivations for the canal was one of national security: it could provide safe passage for shipping, including the Royal Navy, avoiding the threat of privateers when taking a more dangerous (and much longer) route through the

Pentland Firth and around Cape Wrath ('Heritage', 2018). This argument was reiterated in 1842, in an article that lamented the 'dilapidated and neglected state' of the canal. Captain Sir W. E. Parry, appointed to examine the canal in line with the repairs and improvements suggested, asserted that 'in case of war with any nation fitting out fast sailing vessels or steam vessels as privateers, the passage by the Caledonian Canal for merchant vessels would almost wholly supersede that by the Pentland Firth', especially as 'a single efficient man-of-war' would protect the approaches to the canal by sea. He concluded that the canal would 'prove a solid and lasting advantage to the commercial interests of the country' ('Caledonian Canal', 1842).

These six representations of the canal, then, in travelogue, newspaper, and pamphlet, demonstrate the ways in which it was seen as a route through spectacular Scottish scenery for tourists (including highly qualified engineers), a method for protecting British shipping from attack by other nations, international shipping lane, and national transport route. It is only as a tourist route that it is proposed as Scottish; all other uses are folded into a British outlook.

4.5 MANCHESTER SHIP CANAL

According to Forbes, 'in England and Wales the development of inland navigation, which began in 1423 with an act for removing obstructions in the Thames, may be said to have practically ended with the completion of the Manchester Ship Canal in 1894' (1908). He suggests that this canal, begun in 1885 and opened for traffic in May 1895, was 'the most important of the British canals, both commercially and from an engineering point of view' (1908). As the Bridgewater Canal connected mines to Manchester, so the Ship Canal connected Manchester to the coast, making these two waterways fitting markers of an extended canal age. The canal provided access to the city for large ocean-going vessels so, despite being 40 miles from the sea, the canal enabled the Port of Manchester to become the third busiest in Britain. Once work began on the Ship Canal, other British cities began to consider the advantages of similar schemes. The *Hull Daily Mail* reported in 1889 that 'the people of Birmingham [were] seriously considering the advisability of opening out communication by water between the Midlands and the Bristol Channel' and that 'the important town of Sheffield [was] moving in the same direction' ('Ship Canals', *Hull Daily Mail*, 8 March 1889).

An Illustrated History of the Manchester Ship Canal is a contemporaneous celebratory work about the project's planning and delivery, lionizing the men and engineering marvels involved. For instance, there are photographic portraits of the Lord Mayor, Alderman Marshall and of Joseph Leigh, MP for Stockport, and a description of one of the 'great engineering feats' at Ellesmere Port, the building of an embankment over a mile long across the bay. At a price of one penny, a middle-class readership in Manchester could easily be found for this publication, and its readers were exhorted to also obtain *The Visitors' Illustrated Guide to the Ship Canals* and *The Ship Canal Route* (again, for a penny), and subscribe to the monthly penny-illustrated *Ship Canal News*. Photogravure and photo print views were published too: as well as physically looming large in the landscape of the north-west, the meaning of the Ship Canal was produced regionally by these books and guides, which had moved a long way from Hassell's coloured engravings to utilise modern reproduction techniques in emphasising the modernity of the project. In this first example, the *Illustrated History* proposes the canal as a triumph of and benefit to Manchester first and foremost. At a meeting at the Town Hall in 1882, two of Manchester's MPs proposed the importance of the proposed canal as 'essential' to the *district* to find more efficient means of transporting goods and materials between Manchester and surrounding towns and cities. In order to 'push the undertaking' sympathy and support were sought and found in neighbouring towns (Illustrated History, 1894). Daniel Adamson, chair of the Ship Canal's provisional committee when the case went to parliamentary enquiry, is hyperbolically lauded in this text, and described as labouring in his great devotion to the 'social interests and industrial position' of the city of Manchester. In this heroic vision, it is Manchester that opens her gates to the commerce of the world, not Britain (Illustrated History, 1894). This municipal pride was well-founded; when, in 1891, costs for the canal began spiralling, it was the Manchester Corporation which gave generously. The Corporation applied to parliament and was granted the power to borrow £3 million to complete the canal.

By the time the Manchester Ship Canal had opened, British industrial waterways were well established, and the case for such relatively novel infrastructure was no longer needed, so support for the project in this later example was not framed in grand national terms, but in proud regional ones. At a public meeting of working men in the Free Trade Hall in Manchester in November 1882, the *Manchester Courier* reported the chairman's assertion that the canal should be supported by anyone 'who desired to see our commercial supremacy maintained'. The 'our' is 'Manchester's', and the 'all' crosses class groupings; the chairman added that 'it was not a matter affecting specially any one class [with] the bulk of the working classes [understood] that the prosperity of those [mercantile and manufacturing] interests must in the end be their prosperity also'. A later speaker suggested that clothes and food would be cheaper for everyone. The fallacy of these trickle-down economics (to map these hopes in Manchester onto later rhetoric) is borne out by a parliamentary report on poor relief, which noted that many of those coming to the area to find work on the new canal were disappointed: 'Manchester Ship Canal works have caused increase in number of vagrants in Lancashire' (House of Commons, 1888). Mr R. Leake made a contribution via letter, imagining a huge future 'northern metropolis' of the cotton industry with Manchester at its centre; the district, especially Manchester, benefits ('The Proposed Ship Canal to Manchester', 1882).

By 1883, when the parliamentary bill to proceed with the canal had been thrown out, apparently (as the *Manchester Courier* reported it) on a technicality, Alderman Harwood began to see national interests as contrary to those of the city of Manchester. He suggested that many men had invested in the scheme with little hope of seeing a return in their lifetime, but had done so for 'the national weal'. Having proved that the canal would be of benefit to the country, this 'industrial class of people' had been let down by those in Westminster, and the nation would be 'outstripped' by others for its failure to proceed with the ship canal ('Manchester Ship Canal', 1883). Meanwhile, national reporting on the canal was largely limited to its passage through parliament rather than the earth.

4.6 CONCLUSION

In researching King George III's topographical collection, Mercedes Cerón found that maps of canals appeared as a distinct section, 'clearly expected to be part of a working royal collection' because 'improvements in transportation were essential to the success of British manufactures', and civil engineering 'crucial to the prosperity of a commercial nation'. Unlike the canals of France and Germany, maps of which are also in the King's Topographical Collection, the 'construction of British canals relied on the pooling of resources from local landowners and industrialists, rather than on the King and his engineers (Cerón, n.d.). Britain's canals were vast collective political, practical, and commercial undertakings. They required aristocratic political weight, wealthy financial backers, energetic promoters, cooperative landowners, new civil engineers, a huge workforce to build them, and a remodelling of trade and manufacture transport to make use of the new infrastructure. As newspapers and pamphlets lauded the projects, they appealed to a concept of the 'national weal' that cut across class and region. Reading the discourse of nation in this corpus of representations of the waterways, I suggest that many of these texts, written by the powerful and wealthy, argue for the potential of the canals to produce not just national economic benefit, but a cohesiveness born of mutual industrial effort. Fulton's demonstration that Britain followed rather than led in the development of waterways, Stockman's sardonic puncturing of this lofty patriotic ideal, and Alderman Harwood's early identification of an industrial north ignored by Westminster show the limits of nationalist waterways rhetoric, and that pro-canal lobbying was not always successfully articulated via a national vision. A national social hydrological discourse centred on the waterways does not speak with one voice: it shouts in the service of canal promotion, admits disappointment about a lack of support, laughs at self and others, and wonders at the technologically new. Thinking about the waterways in the context of the nation is not just a central aspect of studying historical water management in Britain, it is a generative way of taking a fresh look at how Britain used water technology and transport to create a vision of itself in an extended canal age.

REFERENCES

Anderson B. (2006). Imagined Communities: Reflections on the Origin and Spread of Nationalism, rev. edn, Verso. London and New York.

A Proud Heritage. The Bridgewater Canal. http://www.bridgewatercanal.co.uk/history (accessed 28 March 2018).

Ascott D. E. and Lewis F. (2000). Motives to move: reconstructing individual migration histories in early eighteenth-century Liverpool. In: Migration, Mobility and Modernization, D. Siddle (ed.), Liverpool University Press, Liverpool, pp. 90–118.

Bagwell P. S. (1988). The Transport Revolution. Routledge, London.

Beadon G. (1848). Ten Minutes' Reading of Plain Observations upon Canals and Navigable Rivers. Chapman and Hall, London.

Belchem J. (1999). The Liverpool-Irish enclave. *Immigrants and Minorities*, **18**(2–3), 128–146.

Belchem J. (2005). Priests, publicans and the Irish poor: ethnic enterprise and migrant networks in mid-nineteenth-century Liverpool. *Immigrants and Minorities*, **23**(2–3), 207–231.

Bhabha H. (1994). The Location of Culture. Routledge, London and New York.

Blair J. (ed.). (2008). Waterways and Canal-Building in Medieval England. Oxford University Press, Oxford.

Burton A. (2005). The Canal Builders. Tempus, Stroud.

Caledonian Canal. (1827). Morning Post. 29 December 1827; originally published in *Inverness Courier*. British Library Newspapers. http://tinyurl.galegroup.com/tinyurl/6Hx3E0 (accessed 10 April 2018).

Caledonian Canal. (1840). Caledonian Mercury. 18 July 1840. British Library Newspapers. http://tinyurl.galegroup.com/tinyurl/6J22w6 (accessed 10 April 2018).

Caledonian Canal. (1842). Caledonian Mercury. 14 March 1842. British Library Newspapers. http://tinyurl.galegroup.com/tinyurl/6J2fG1 (accessed 10 Apr. 2018).

Cerón M. (n.d.). Mapping Europe's waterways in George III's topographical collection. In: Picturing Places. British Library, London. https://www.bl.uk/picturing-places/articles/mapping-europes-waterways-in-george-iiis-topographical-collection (accessed 29 October 2020).

Constantine M.-A. (n.d.). John Hassell, Wales and the 'industrial picturesque'. In: Picturing Places. British Library, London. https://www.bl.uk/picturing-places/articles/john-hassell-wales-and-the-industrial-picturesque (accessed 29 October 2020).

Fairclough K. R. (2009). 'Egerton, Francis, third duke of Bridgewater (1736–1803), canal promoter and colliery owner.' Oxford Dictionary of National Biography. Oxford University Press. See https://www.oxforddnb.com/view/10.1093/ref:odnb/9780198614128.001.0001/odnb-9780198614128-e-8584

Forbes U. A. (1908). The inland waterways of Great Britain and the plans under consideration for their improvement. *Annals of the American Academy of Political and Social Science*, **31**, 228–245.

Fulton R. (1796). A Treatise on the Improvement of Canal Navigation. I. and J. Taylor, London.

Great Exhibition of the North. (2018). https://getnorth2018.com/ (accessed 9 May 2018).

Green R. (2017). Almost £1 m of Lottery funding for 'Kick the Dust' to involve young people in Yorkshire's heritage. IVE. https://weareive.org/almost-1m-lottery-funding-for-kick-the-dust-to-involve-more-young-people-in-yorkshires-heritage/ (accessed 10 May 2018).

Hassell J. (1819) Tour of the Grand Junction. Published Privately, London.

Hay D. and Rogers N. (1997). Eighteenth-Century English Society. Oxford University Press, Oxford.

Heritage. (2018). Scottish Canals. https://www.scottishcanals.co.uk/heritage/caledonian-canal/ (accessed 9 May 2018).

Hibbard J. (1796). Thoughts (Submitted to Censure) on the Great Utility of a Circular and Other Inland Canal Navigation in Britain. n. pub., London. Available via wellcomelibrary.org (accessed 10 May 2018).

Hill S. (2017). The socio-cultural impact of the War of American Independence on Liverpool, 1775–83. *Journal for Eighteenth-Century Studies*, **40**(2), 163–180.

House of Commons. (1888). Index to the Report from the Select Committee of the House of Lords on Poor Relief. Hansard, London. https://archive.org/details/b24399504 (accessed 10 May 2018).

Illustrated History of the Manchester Ship Canal. (1894). Heywood, Manchester. Online. http://access.bl.uk/item/viewer/ark:/81055/vdc_000000002F16 (accessed 8 April 2018).

In and Out of Place. The Telegraph. 7 October 1795. Available via the British Library 17th and 18th Century Burney Collection. http://tinyurl.galegroup.com/tinyurl/6BVT95 (accessed 23 March 2018).

Lee R. (2012). Divided loyalties? In-migration, ethnicity and identity: the integration of German merchants in nineteenth-century Liverpool. *Business History*, **54**(2), 117–153.

Linton J. (2010). What is Water? The History of a Modern Abstraction. UBC Press, Vancouver and Toronto.

Manchester Ship Canal. (1883). Manchester Courier and Lancashire General Advertiser. 29 August 1883. British Library Newspapers. http://tinyurl.galegroup.com/tinyurl/6NkqFX (accessed 20 April 2018).

McCahill M. (1976). Peers, patronage, and the Industrial Revolution, 1760–1800. *Journal of British Studies*, **16**(1), 84–107.

McIvor L. (2015). Canals: The Making of a Nation. BBC Books, London.

Morecroft E. (2017). Antislavery, elite men, and the 'voice of the British nation:' c. 1790–1860. *History Compass*, **15**(5), e12374.

Petticrew I. and Austin W. (2012). A Highway Laid with Water: An Account of the Grand Junction Canal, 1792–1928. http://gerald-massey.org.uk/canal/index.htm (accessed 14 April 2018).

Purdue L., Salomon F., Berger J.-F. and Goiran J.-P. (2015). Canal through time: towards a multidisciplinary and holistic study of water systems? *Water History*, **7**, 1–9.

Samuel S. (1868). The Life of George Stephenson and of his Son Robert Stephenson. Harper, New York. https://archive.org/details/lifeofgeorgestep00smiluoft (accessed 10 May 2018).

Stockman B. P. (1861). A Trip Through the Caledonian Canal. Privately printed, London. See http://access.bl.uk/item/viewer/ark:/81055/vdc_000000058D10 http://access.bl.uk/item/viewer/ark:/81055/vdc_000000058D10#?c=0&m=0&s=0&cv=0&xywh=-626%2C-141%2C2803%2C2810

Ship Canals. Hull Daily Mail. 8 March 1889. British Library Newspapers. http://tinyurl.galegroup.com/tinyurl/6VDfp2 (accessed 9 May 2018).
Smiles S. (1868). The Life of George Stephenson and of his Son Robert Stephenson. Harper, New York. https://archive.org/details/lifeofgeorgestep00smiluoft (accessed 10 May 2018).
The Great Commercial Canal. True Briton. 23 November 1795.
The Great Commercial Canal. The Times. 24 December 1795. The Times Digital Archive. http://tinyurl.galegroup.com/tinyurl/6VDdA8 (accessed 9 May 2018).
The Proposed Ship Canal to Manchester. (1882). Manchester Courier and Lancashire General Advertiser. 27 September 1882. British Library Newspapers. http://tinyurl.galegroup.com/tinyurl/6Nn626 (accessed 20 April 2018).
Thompson C. (n.d.). The picturesque at home and abroad. In: Picturing Places, British Library, London. https://www.bl.uk/picturing-places/articles/the-picturesque-at-home-and-abroad (accessed 29 October 2020).
Vedmore M. (2017). The Canoeist and the Royal Canoe Club. In: Canal Book Collector. http://canalbookcollector.blogspot.co.uk (accessed 2 April 2018).
Young A. (1770). A Six Months Tour Through the North of England. Strahan, London. https://www.bl.uk/collection-items/a-six-months-tour-through-the-north-of-england (accessed 10 May 2018).

Chapter 5

A Few water fountains in Tampere, Finland: Cultural attestation of water management

Petri S. Juuti and Riikka Rajala

ABSTRACT

This chapter describes how water influenced the growth of Tampere on the basis of four major water fountains related to the industrial heritage of the city. Tampere, currently Finland's third biggest city, with a population of over 220,000, and the largest inland city in the Nordic countries was founded in 1779. For many years, Tampere was the centre of Finnish industry. The history and development of Tampere have been largely shaped by issues related to water and water management. The city waterworks were established in which connection the industrialist Wilhelm von Nottbeck donated the fountain in the Central Square in 1882. A fountain depicting a swan and a boy and another with a lion theme are also connected to the industrialist von Nottbeck. Each fountain has its own history to tell viewers. Water fountains are probably at their best when they represent and reflect local city culture and heritage as these examples from Tampere.

Key words: Water fountains, water history, Tampere, industrial history

5.1 BACKGROUND: INDUSTRIAL HERITAGE OF TAMPERE

Finland is unique in Europe as to its water resources. The country has some 56,000 lakes with a minimum area of one hectare and all together about 200,000 lakes.

Editor: Xiao Yun Zheng
doi: 10.2166/9781789062045_0069

Groundwater occurs in alluvial eskers formed during the ice ages, the last of which ended some 10,000 years ago. Areas lower than 50–60 metres above sea level have problems with water quality for geological reasons. In such areas bigger cities use surface water for their water supply or acquire their raw water from sources further away. Nowadays some 60% of the people use natural groundwater or artificial recharge.

Tampere, currently Finland's third biggest city and the largest inland city in the Nordic countries with a population of over 220,000, was founded in 1779. For many years, Tampere was the centre of Finnish industry which is also reflected in its nickname "Manse" which derives from Manchester, the traditional English industrial centre. Many industries and factories established in Tampere between the 1830s to the 1950s were among the first of their type in Finland. In later years several high-tech innovation "firsts" have also been launched in Tampere.

The history and development of Tampere have been largely shaped by issues related to water and water management. The city was established along the rapids that cut their way through a glacio-fluvial esker between two large lakes some 7,500 years ago. The availability of water power and lakes suitable for waterborne transport made the Tammerkoski Rapids and its banks a favourable location for a growing town (Katko & Juuti, 2007). This chapter describes how water influenced the growth of the city on the basis of four major water fountains related to the industrial heritage of Tampere.

The first municipal water pumping system in Tampere was built in 1835. It had a German-made iron pump upstream of the Tammerkoski Rapids which pumped water through a wooden pipe to a well in the market square. The system was a failure, but it was only after the fire of 1865 that the piped-water issue was taken up again. The industrialist Wilhelm von Nottbeck suggested to the City Board that a privately owned waterworks be constructed for the city. Von Nottbeck was the owner of the biggest industrial enterprise in Tampere, the Finlayson Mill. Since the proposed privately owned waterworks would have meant a quite big financial risk to the city, the Board rejected the proposal and decided that the city would build its own water system (Juuti & Katko, 1999; Juuti, 2001).

Subsequently, the city sought expert knowledge and contractors from the private sector, such as Mr Malakias Pasi in 1874. The following year the City Board decided "to submit the piped-water issue to a committee for examination". This meant the start of a new era in the history of water supply in Tampere: ever since all related plans and projects have been initiated by the city.

The demand for improved water and sewerage services (WSS) was created by several factors based on different local conditions. Such factors could include simply the needs of WSS business, fire protection, water quantity and quality problems, public health; and later environmental protection; other influences include industrial or tourism water use, regional focus, or various combinations of these. The business motive was the cornerstone of the first private proposal in the mid-1800s as well as some 130 to 140 years later. Fire protection was particularly important in areas with wooden houses. Although in many cases public health requirements were the major driver, water quality and quantity also had an impact. In some cases productive use rather than community needs has been the major driver. Water fountains are linked to all of these aspects: now they are important for aesthetics, but in the first phase they were also used as a source of drinking water for domestic animals, especially for horses on market days, and people passing by. Later on, they served as an oasis of peace, a place to relax for a moment in the hectic city centre (Juuti & Katko, 2005). The fountains were also an outcome of water culture and also played a cultural role in water management.

5.1.1 The central square fountain

In 1882 the city water works of Tampere were established in which connection the industrialist von Nottbeck donated the fountain in the Central Square (Figure 5.1). It was the harbinger of a new era – an era of water in many senses (Juuti, 2001; Juuti & Katko 1999; Tampere City Records Office, 1874).

The fountain was donated particularly to the women of the town to provide them with household water. The cast iron fountain was made in Potsdam, German Empire. Under Wilhelm von Nottbeck, the Finlayson Cotton Mill grew into

Figure 5.1 Central square fountain that has operated since 1882 in front of the city hall of tampere. (Photo: P. juuti).

Finland's first large-scale industrial enterprise. The fountain was the end point of a water pipe. It was a place where water could be drawn freely for household use and animals could be watered. The fountain stands right in front of the old City Hall of Tampere, currently the major venue for the city's public relations events.

The fountain has been in continuous operation since 1882, except for periods of restoration. In 2004 a 4.5 metre high glass dome-type cover was installed to protect the fountain, which represents the city's cultural heritage, in winter. The cover is normally removed in April or May and put back in place in October. The cover was manufactured by Tambest Ltd and designed by the architect Juha Ryösä. The high-tech glass structure is made of 6 mm safety glass laminated to 6 mm Pilkington self-cleaning glass. The lower and upper parts of the dome have sand-blasted decorations designed by Sauli Iso-Lähteenmäki. The dome is illuminated during dark winter evenings and nights (Hynynen *et al.*, 2012).

This fountain was directly connected to the city water works from the beginning of its long history. In 1881, the governor of the province ordered a water works to be built in the city of Tampere due to recurring fires and the poor quality of well water. The same year construction began on a gravity water pipe from Mältinranta, upstream of the rapids, to a market square in the southern part of the city – the pipe also had several connections. The contractor was R. Huber who supervised the construction of Helsinki (the capital city of Finland) Water Works the previous decade, and also operated the system for a few years. The pressure of the system was low but technically and administratively it operated much like a modern water works (Juuti, 2001).

Several tall apartment blocks were erected in the city centre towards the end of the nineteenth century. Then it was realised that water pressure was too low to carry water to the top floors. Therefore, the city council requested that C. Hausen, the engineer of the Helsinki Water Works, design a new high-pressure facility. The facility which was operated according to quite modern principles was completed in 1898. It had an underground tank high up in the Pyynikki Ridge which is still in use a hundred years later providing sufficient pressure for the core city.

5.1.2 The fountain in Wilhelm von Nottbeck park

The second and third fountains described here are also related to the industrialist von Nottbeck. In 1995 the park around the Finlayson Palace came into the city's possession and was renamed the Wilhelm von Nottbeck Park. In 1820 – some 175 years earlier – a Crown distillery and gardens were situated here.

The original park was created in the 1840s when Wilhelm von Nottbeck was the owner of the Finlayson Mill. The park was designed in the English style with flower beds, curved walks, three islands with connecting bridges and five gazebos. The channels between the islands were filled between 1922 and 1923, which altered the shoreline and the shape of the entire park. No plan or map of the original park has been found, but it is likely that the park was created by skilled master gardeners.

During the time the von Nottbeck family lived in the palatial house, the park had four large greenhouses where seeds and bulbs were brought in to bloom. Many varieties of fruit were also grown in the greenhouses, such as watermelons, pears and peaches. Grapes and pineapples were grown in separate greenhouses, and the grape crop, in particular, was very large. The greenhouses were dismantled in 1978 to make way for a parking area – a new priority. In 1899 the old wooden main building was replaced by a new palace in the neo-Renaissance style, designed by the City Architect Lambert Petterson. Today the renovated palace functions as a high-quality restaurant.

The bronze statue of an eagle on the Kotkankallio rock is regarded as the oldest structure in the park. One wooden gazebo, called the Chinese Parasol, remains in the northwestern corner of the park. The palace is fronted by a bronze fountain displaying a swan and a boy (Figure 5.2). Another fountain, with a lion theme, is located north of the main building (Figure 5.3).

5.1.3 The fountains of Hämeenpuisto Boulevard Park

Our fourth fountain stands in what today constitutes the western part of the city, west of the north–south Tammerkoski rapids. The kilometre-long Hämeenpuisto Boulevard Park was built in 1909 in the north–south direction. Its original

Figure 5.2 Fountain: boy and swan next to finlayson palace (Photo: P. juuti).

Figure 5.3 Lions fountain next to finlayson palace (Photo: P. juuti).

purpose was to act as a firewall between the two sections of the city, since all houses were wooden at that time and even much later. The park has numerous benches, where strollers can take a breather and just watch people passing by.

Large fountains stand at both ends of the Boulevard. The Näsi Park at the northern end was created in the early 20th century on almost bare rock using earth dredged from the nearby harbour, and has a total area of about 6 hectares. The park was planned by the City Gardener Onni Karsten. Only spindly pines and junipers used to grow on the rock, but now the park boasts hundreds of trees and some 500 perennials, including rare imported species. The fountain at the northern end is the first public work of art in Tampere, dating from 1913. It was designed in the national romantic style by Emil Wikström and donated by the wealthy merchant, Nikolai Tirkkonen (Figure 5.4).

The fountain comprises three statues. One depicts the maiden Pohjanneito spinning gold thread, while the other two on a lower level symbolise the cycle of life in the form of a grandfather and his grandson standing by a water wheel, and the grandmother teaching her granddaughter how to weave. Thus they reflect the industrial heritage of the city. In October 2008 specially designed lights were installed in the fountain to illuminate it.

Figure 5.4 Tirkkonen fountain (Photo: P. juuti).

At the other end of Hämeenpuisto Boulevard we find the South Park, created in two phases between 1915–1917 and 1928–1929. It was designed by City Engineer K. Vaaramäki, and its construction was led by City Gardener Onni Karsten. As late as 1897 Russian soldiers had vegetable gardens and root [earth] cellars in the park, the reason being that Finland was a Grand Duchy of Russia at that time. In 1918, during the civil war fought after Finland gained her independence in 1917, potatoes and swedes were also grown there.

In the 1920s the South Park had a classical-style stage with roses growing around it. Today a granite monument and fountain honouring the co-operative movement, sculpted by Wäinö Aaltonen in 1950, stand in its place. The two large fountains at the northern and southern ends of the Boulevard Park represent two different ideologies – once so far apart, but now co-existing in harmony (Figure 5.5).

Figure 5.5 The co-operative fountain (Photo: P. juuti).

Figure 5.6 Modern fountain next to the library. (Photo: P. juuti).

There is also a fairly recently erected fountain consisting of a granite ball rotating on a layer of water in the middle of the Boulevard Park, next to the city main library called Metso (capercaillie in English). It is also protected by a glass cover in winter (Figure 5.6).

5.2 DISCUSSION AND CONCLUSIONS

All in all, the Tampere City Water and Sewage Works is an example of a local government-owned public utility that has been, and will continue to be, capable of providing services at reasonable cost. From the very beginning public/private partnership has taken place: the core competence is taken care of by the municipality-owned utility, while supporting services are bought from the private sector based on competition. The utility has decisively improved the city's fire safety, hygiene and health conditions, and the quality of the city environment. It has also been central in enhancing the operating conditions of industry and commerce. Although many facilities of the works are hidden underground, we all daily come into contact with its key products: potable water, wastewater and cleaner water bodies. Luckily, the beautiful water fountains are great remainders of the importance of water supply system and they are visible to all.

Water fountains are probably at their best when they represent and reflect local city culture and heritage like the fountains in Tampere, described in this account. They are all closely connected to local industrial heritage and traditions – even their designs reflect historical linkages in many ways. They are not only beautiful and impressive, but also tell a story of industrial revolution in Tampere. The more recent illumination of the fountains has only added to their beauty and appeal.

The industrial heritage of Tampere is reflected by water in many ways. As the final touches are being put on this chapter, the city is preparing a proposal for the

Tammerkoski Rapids flowing through the city centre, and the old factories along its banks, to be placed on the UNESCO World Heritage List (Järvi, 2007).

Even in a four-season climate, winter does not have to prevent the operation of water fountains the year round. After all, greenhouses operated in Tampere throughout the year already about 150 years ago. Water fountains, especially those in the public spaces of the city, reflect different aspects of industrial heritage and, indirectly, its other cultural traditions and evolution.

ACKNOWLEDGEMENTS

We thank VEPATUKI research cluster and Prof. Zheng Xiao Yun from support.

REFERENCES

Hynynen A., Juuti P. and Katko T. (eds) 2012. Water fountains in the worldscape. International Water History Association and KehräMedia Inc.

Järvi M. (2007). The history and future of tammerkoski and its banks – a UNESCO world heritage?. In: 2007. Pasts and Futures of Water. Abstract book. 5th IWHA 2007 Conference, I. Hautamäki, P. Juuti, T. Katko, R. Rajala and E. Vinnari (eds), 13–17 June, 2007, Tampere, pp. 114–115.

Juuti P. (2001). Kaupunki ja vesi. Tampereen vesihuollon ympäristöhistoria 1835–1921. Doctoral dissertation. University of Tampere. (City and Water. Environmental history of Tampere 1835–1921). Summary in English. http://acta.uta.fi/pdf/951-44-5232-1.pdf

Juuti P. and Katko T. (1999). From polluted to swimmable water: tampere city water works 1835–1998. *Water International*, **24**, 204–210.

Juuti P. and Katko T. (eds.) (2005) Water, Time and European Cities. www.watertime.net

Katko T. and Juuti S. (2007). Watering the city of tampere from the mid-1800s to the 21st century. Tampere Water & International Water History Association. www.tampere.fi/tiedostot/5pxlRx6kV/Watering_the_City_of_Tampere.pdf

Tampere City Records Office. (1874). Minutes of City Administrative Court as of 28.9.1874

Section 3

The Water Culture in Indigenous Societies

Chapter 6

Learning from indigenous water ethics

David Groenfeldt

ABSTRACT

Even as Indigenous cultural lifeways are gaining newfound respect, as evidenced by the UN Declaration of the Rights of Indigenous Peoples (2007), the very real threats to traditional Indigenous land and waters, and indeed individual physical safety, continue unabated. This chapter focuses on Indigenous efforts to protect culturally important water ecosystems and identifies lessons that can be usefully applied to the modern challenges of sustainable water governance. Water ethics provides a common framework for articulating water's ontological significance for Indigenous communities in contrast to the role ascribed to water in Western cultural contexts. For Indigenous societies, rivers are living beings that nourish human communities and that humans have a responsibility to protect. It is a profoundly Western notion that people have a moral duty to put rivers to work for economic benefits such as irrigation or hydropower. Ontologies matter: The ethical response to a living river is to protect its life, while the ethical response to a river viewed as a lifeless economic resource is to extract as much benefit from that resource as possible. Both Indigenous and non-Indigenous societies practice what they each consider to be ethical water management. When the sustainability of water resources is considered, however, Indigenous ontologies have an adaptive edge. Clearly there are potential lessons for sustainable development!

Key words: Indigenous, Water ethics, sustainability, ontologies

Editor: Xiao Yun Zheng
doi: 10.2166/9781789062045_0081

Indigenous Peoples weave natural resources – water, land, crops, and forests – into their cultural identities. The interplay of culture and nature has evolved over centuries and millennia, binding their fates inextricably together. The Egyptian civilization was literally born of the Nile River; the Hopi People in the Southwestern United States trace their ancestry from water beings. There is an elemental connection of relatedness that informs Indigenous Peoples' economic exploitation of nature. Hunting can take the form of a ritualistic conversation with the animals being hunted, so that even killing in this context has the effect of further connecting the human and natural worlds. In such a world, the welfare of hunted animals is viewed as an ethical responsibility of local people, just as contemporary Indigenous Water Protectors in the United States feel a responsibility for the Missouri River, threatened by contamination from oil pipelines. Indigenous ecological knowledge constitutes a cultural heritage that can inform our present and our future, and not only the past (Suzuki & Knudtson, 1992).

Yet even as the relevance of Indigenous Traditional Ecological Knowledge is becoming more widely acknowledged, the cultural integrity of Indigenous communities continues to be sacrificed in favor of dams, mines, and other forms of conventional economic development. The discredited economic logic of the 20th century, which the UN Sustainable Development Goals (SDGs) are seeking to overturn, continues to guide real-world decisions. There is a terrible disconnect between what the world is doing (e.g., burning fossil fuels) and what the world is saying it wants to be doing (e.g., reduce global warming). The future we are actually creating is almost the antithesis of 'The World We Want' that is described in UN sustainability meetings. This disconnect is nowhere more apparent than in the water sector, where freshwater quality continues a relentless trajectory of degradation even as river restoration science becomes ever more sophisticated (Arthington *et al.*, 2018).

The prevailing neoliberal water ethic that views water as an economic resource to be exploited for production and profit is a manifestation of contemporary water culture. According to this ethic, water managers have a responsibility to extract as much water as possible from nature, in order to generate economic value from that water. These materialistic values favor short-term profits over long-term ecosystem sustainability and lead almost inevitably to the degradation of water resources. This is no longer a controversial assertion; the danger of basing water policies on neoliberal economic principles is widely acknowledged. The UN High Level Panel on Water adopted the Bellagio Principles on Water Values to highlight the need for new ways of valuing water beyond monetization (HLPW, 2017). Solving the water crisis will require a re-valuing of water to incorporate core principles of environmental sustainability, social justice, and transparent governance.

Where can we find the value principles to inform a new ethics of water that meets the tests of sustainability, equity, and justice? There are many models and schools of thought to choose from; we can find diverse value perspectives within both Western and Oriental philosophical traditions which can guide us to become better stewards of water. But in my view the most compelling guidance for transitioning away from our current destructive relationship with the natural world is to be found within the value systems of contemporary Indigenous Peoples. The very people who are most threatened by extractive industries have much to teach us about coexisting with the natural world. By emulating Indigenous value systems, we can learn how to rebalance our out-of-control exploitation of water. We can, and in order to meet the goal of sustainable development, we *must*, rebalance water management to emphasize water protection rather than water extraction.

6.1 WATER ETHICS AS AN ASPECT OF WATER CULTURE

Values refer to 'standards or criteria to guide not only action but also judgment, choice, attitude, evaluation' etc. (Rokeach, 2000, p. 2). Ethics refers to the principles we adopt in order to apply our values within the complicated settings of actual behavior. Both our values about how water should be used and how rivers should be managed, and the ethical principles that guide our actual water behavior, can be considered aspects of water culture (Groenfeldt, 2019, pp. 6–7). Ethics evolve as values change. We can observe this dynamic in the way water policies are changing to reflect greater concern about protecting aquatic biodiversity by maintaining at least minimum environmental flows. As our values shift towards greater concern for environmental sustainability, the ethical principles which define ethical behavior also shift. Maintaining environmental flow has become expected behavior in water governance and is mandated in the national water laws of Australia, South Africa, and (somewhat indirectly) the EU Water Framework Directive (Howarth, 2018).

The rationale given for environmental flow policies, however, is not usually an ethical argument, that it is the right thing to do. Instead the environmentally friendly policies are justified with an economic argument, that maintaining ecological function makes good *economic* sense. The economic value of the ecological services of a flowing river – including the (economic) benefits of recharging riparian aquifers, supporting riverine fisheries, protecting river-side real estate values, etc. – is greater than the economic benefits of fully using and depleting the river. From a water resources perspective, maintaining environmental flow is a means for protecting the integrity of the water resource.

Additional benefits from a flowing river could also be cited such as social equity (All stakeholders have a right to a flowing river) and cultural/spiritual values (We enjoy seeing the river flowing). These 'soft' arguments in favor of environmental flow requirements may be fully consistent with stakeholder values, but are in a sense superfluous. Since water is categorized as a resource, an economic argument is adequate, and it is already ethical: It would be unethical to not protect the sustainability of the water resource.

I am presenting this example to show that (a) Water ethics is a key dimension of water culture, and (b) Within the paradigm of Integrated Water Resources Management (IWRM) where water is viewed as a resource, the economic values of water enjoy a privileged position. Thus the ethics that matter most are the economic ethics of using resources wisely. Caring for rivers because of inherent rights of rivers to remain healthy lies outside the 'ontology' of IWRM. Legal efforts to recognize rivers as legal persons, following the case of New Zealand granting legal personhood to the Wanganui River (Iorns Magallanes, 2018) have met solid resistance. The enabling legal and moral environment does not seem conducive to valuing rivers for their own existence in a way analogous to human rights laws enacted to protect people. Indigenous ontologies of water, however, provide space to Nature and to lakes and rivers. What we can learn from Indigenous understandings of what water is, and what rivers are, is an alternative worldview where rivers are relatives and we humans have a responsibility to respect our water relatives. What are the ethical principles that would support a world where rivers are our relatives? (Photo 1)

Photo 1 Is India's Mahanadi River a 'resource' or a relative?.

6.2 INDIGENOUS VIEWS OF WATER

Tapping into the knowledge and philosophical perspectives of Indigenous Peoples is both a utilitarian strategy (for crafting more effective sustainability interventions) and a moral responsibility. We have a 'duty to learn' about Indigenous concepts lest we inadvertently undermine the very ideas that are so important to preserve. Chief Justice Lance S.G. Finch of the British Columbia (Canada) Court of Appeal defines the settler-colonialists' 'duty to learn' as a first step to accommodating cultural diversity (Finch, 2012): 'How can we make space within the legal landscape for Indigenous legal orders? The answer depends, at least in part, on an inversion of the question: a crucial part of this process must be to find space for ourselves, as strangers and newcomers, within the Indigenous legal orders themselves.' In other words, we have an obligation, a duty, to learn how Indigenous Peoples view water and water ecosystems, lest we repeat the mistakes of 19th century missionaries who sought to convert Indigenous Peoples to a Western belief system, which today is implicated as contributing to runaway climate change. Better understanding of and appreciation for the water ethics of Indigenous Peoples can help us rebalance our unsustainable water habits.

Given the many thousands of distinct Indigenous cultural groups in the world, we must be cautious about overgeneralizing a uniform Indigenous world view about water. Yet when representatives of diverse Indigenous groups convene in international water conferences, they are able to find agreement about certain basic principles. An illustration of this agreement-making capacity is the Indigenous Peoples Kyoto Water Declaration formulated by Indigenous participants at the 3rd World Water Forum in Kyoto, Japan in 2003. The key drafters were Tom Goldtooth, the founder and president of Indigenous Environmental Network (http://www.ienearth.org), and Victoria Tauli-Cruz, who at the time was Executive Director of Tebtebba (and later became the UN Special Rapporteur on The Rights of Indigenous Peoples). The Kyoto Declaration was communicated to the World Water Forum by a march through the conference center, followed by a press conference. Later the Declaration was posted on various websites and was also included in the UNESCO publication, *Water and Indigenous Peoples* (Boelens *et al.*, 2006).

The Indigenous Peoples Kyoto Declaration on Water outlines two fundamental themes of water ethics: (1) The relationship that Indigenous Peoples have with respect to water, and (2) The rights of Indigenous Peoples to water access and self-determination about how water should be used and protected.

6.2.1 Relationship to water

The first three statements of the Declaration describe how Indigenous Peoples are related to water:

(1) We, the Indigenous Peoples from all parts of the world assembled here, reaffirm our relationship to Mother Earth and responsibility to future

generations to raise our voices in solidarity to speak for the protection of water. We were placed in a sacred manner on this earth, each in our own sacred and traditional lands and territories to care for all of creation and to care for water.

(2) We recognize, honor and respect water as sacred and sustains all life. Our traditional knowledge, laws and ways of life teach us to be responsible in caring for this sacred gift that connects all life.

(3) Our relationship with our lands, territories and water is the fundamental physical cultural and spiritual basis for our existence. This relationship to our Mother Earth requires us to conserve our freshwaters and oceans for the survival of present and future generations. We assert our role as caretakers with rights and responsibilities to defend and ensure the protection, availability and purity of water. We stand united to follow and implement our knowledge and traditional laws and exercise our right of self-determination to preserve water, and to preserve life' (ibid, p. 176)

Indigenous Peoples of Australia view water an integral part of the world created by ancestral beings during what is colloquially known as the Dreaming (Jackson & Barber, 2016). The land and the waterscapes are understood to be living entities that are responsive to human actions and human behavior. People place value on their relationships to the landscape, just as they do to one another. This is the meaning of the expression heard in UN speeches by Indigenous Leaders referring to 'All our relations.' Those relationships are with other humans, animals, the landscape, and the waterscapes, and imply both rights (the right to be acknowledged as a relative) and responsibilities (to protect the land, water, and associated plants and animals). There is also a responsibility for understanding nature as a prerequisite to making use of resources: 'Possession of knowledge of the environment, its natural features and vitally, its spiritual dimensions, is a prerequisite to exercising rights to land and water' (Ibid).

6.2.2 Right to water and self determination

The Kyoto Declaration also highlights the rights of Indigenous Peoples to use and enjoy their customary water resources and their broader cultural rights to choose their way of life. These rights were further clarified by the UN General Assembly in the 2007 Declaration on the Rights of Indigenous Peoples, as discussed later in this paper. In the words of the 2003 Kyoto Declaration:

(9) We Indigenous Peoples have the right to self-determination. By virtue of that right we have the right to freely exercise full authority and control of our natural resources including water. We also refer to our right of permanent sovereignty over our natural resources, including water.

(10) Self-determination for Indigenous Peoples includes the right to control our institutions, territories, resources, social orders, and cultures without external domination or interference.

(11) Self-determination includes the practice of our cultural and spiritual relationships with water, and the exercise of authority to govern, use, manage, regulate, recover, conserve, enhance and renew our water sources, without interference'. (Boelens *et al.*, 2006)

6.2.3 Water's meanings

In her analysis of what water means for the Carcross/Tagish First Nation in Canada, Eleanor Hayman (Hayman *et al.*, 2017a) identifies three distinct but synergistic dimensions of the Indigenous water paradigm: (1) Water as metaphor, (2) Water as map, and (3) Water as responsibility.

Water as metaphor: Water serves as a multipurpose metaphor for living one's life in the sense that everything is connected (my life with yours; our human lives with the natural world) and the concept of flow in the sense of a circulation of virtues, or borrowing from Eastern religions, the law of karma. In these senses water provides a useful and ubiquitous metaphor that is good to think with (Chen *et al.*, 2013; Cruikshank, 2012).

Water as map: Water bodies – rivers, lakes, wetlands, springs, or (in the Yukon region especially), glaciers, are ready-made features that can orient direction both physically (north/south, upslope/downslope) and culturally, identified and remembered through story (Basso, 1996). Deep mapping 'intentionally creates space for ontological difference [and] gives voice to aqua-centric wisdom' (Hayman *et al.*, 2017b, p. 237) through documenting the stories linked to water bodies.

Water as responsibility: Indigenous cultures recognize a responsibility to water in the form of respect (through prayers, ceremonies, and storytelling) and protection in the sense of according the category of 'sacred' to water bodies and treating them, and speaking about them, in respectful ways. Respecting and protecting constitute a deep ethical responsibility to water which was displayed on a world stage at the Standing Rock protests in South Dakota, USA during 2016. Native protesters called themselves 'Water Protectors', protecting both the land and water from the oil pipeline being constructed across their territory, and under the Missouri River (Veilleux, 2017). Though not successful in blocking the pipeline, which was supported by an overwhelming coalition of state and local government, the US Army Corps of Engineers, and the pipeline company (supported by heavily armed security forces), the protests clearly illustrated Indigenous values of responsibility to the land and water.

Given the stark contrast between the cultural paradigm of water in Indigenous societies vs. the cultural paradigm of water espoused by US Army Corps of Engineers and the oil companies, how can the two worlds communicate? There are deep conceptual challenges that need to be addressed, having to do with colonization and the hegemony of conventional paradigms of water and development. We need to 'decolonialize' our minds (Porter, 2010) in order to understand Indigenous concepts of what water is, and how to establish an ethical relationship with water. We need to overcome own cultural biases and make space in our minds for other ontologies.

6.2.4 Ontologies of water

'Ontology' refers to fundamental beliefs about the nature of being and existence. Ontologies of water refer to beliefs about the nature of water and bodies of water, such as a river. These issues of philosophical belief take on very practical relevance when something drastic is proposed for the river, such as a dam or a mine that could cause irreparable harm. The success of Western colonialism and the hegemony of Western concepts of economic development were facilitated by political, economic, and military power, of course, but this should not blind us to other less obvious ways that colonial cultures imposed a colonial ontology of water. The colonial impact on Indigenous ontologies was perhaps even more devastating than the physical conquest of territory and governance institutions. Indigenous concepts of nature were violated and continued to be challenged through the process of economic development assistance. Just as Christian missionaries sought to replace Indigenous religions, World Bank 'missions' offered financial and technical assistance to reform key economic sectors including water and agriculture. I refer to the dynamic of Western experts reforming and literally reframing the water sector as a process of 'semiotic hegemony' (Groenfeldt, 2016). Not only are new policies forcibly introduced as conditions of financial assistance, but a new ontology of water comes along with the funding.

Indigenous ontologies are rarely taken seriously if, as they often do, they run counter to Western development paradigms. Yates *et al.* (2017, p.797) suggest that we need '…to take seriously the possibility and politics of a multiplicity of water-related worlds, highlighting multiple water realities and ways of being-with-water, not just different perceptions of or knowledge systems tied to water's (singular) material existence.'

An example of water ontologies in practice is the question of whether a river is a living being or simply a flow of lifeless water. At the 2006 World Water Forum in Mexico City, I co-organized a session about indigenous perspectives on water, along with Tom Goldtooth, of Indigenous Environmental Network. We titled our session, 'Is Water Alive? Indigenous Understandings of Water.' The professional water experts who dominated the World Water Forum had no doubts about the answer to the question we were posing. Of course rivers are not alive! And the Indigenous speakers in our session also had no doubts about the correct answer: Of course rivers are alive; they have agency and can choose how they respond to people. This understanding constitutes the basis for managing water through ceremonies and rituals, as well as through physical practices such as diverting water into irrigation canals. What non-indigenous settlers see as underutilized rivers (because they are not diverted) are perhaps already being managed with prayers and songs which express gratitude for the river for simply being a river.

When different ontologies of water coexist in a single river basin, how can integrated management decisions be undertaken? Cultural differences in worldview, which translate into different value preferences such as whether a proposed dam would be a benefit or a curse, cannot be resolved by more hydrological data or more economic valuation studies. Such deep conflicts of worldview constitute 'ontological disjunctures' (Yates *et al.*, 2017) which require not only intellectual understanding to bridge, but also an emotional appreciation of the other's perspective.

The solution to the challenge of bridging diverse ontologies is to embrace cultural diversity and acknowledge the right to be different, to have different values, and to express different behaviors. But there is a prerequisite to peaceful coexistence of divergent cultures. There must be a willingness to look first for value synergies and to set aside value conflicts, to favor cooperation over conflict. Successful mediation begins with good will on all sides from which appreciation of the other's position and respect for the other's values can lead to trust and eventually cooperation. Practical outcomes acceptable to all parties will be supported for different value reasons. No party receives everything it wants, but the core values of each party are respected and no party suffers egregiously (Wolf, 2017).

6.3 CREATING SPACE FOR DIVERSE WATER ONTOLOGIES

The prevailing paradigm of modern global civilization, whether in its Western (Euro-American), Eastern (Asia), or Southern (Africa and South America) manifestations assumes that traditional Indigenous worldviews cannot possibly persist once their adherents taste the modern tree of rational materialism. Even as the powerful object lesson of climate change undermines the legitimacy of the global economic juggernaut, we remain reluctant to concede the relevance of spiritually-based Indigenous ontologies. With regard to water resources, this same hubris has blinded us to the potential benefits of Indigenous ontologies of water in solving the challenges of water security and sustainability. The solutions to the water crisis that the West continues to put forward, based on some variant of IWRM, are not really working. The world's water resources continue to be degraded in service to short-sighted economic development (IFPRI and Veolia, 2015).

It is not my intent to disparage the role of science or economics in driving water policies, but rather to suggest that the concept of cultural diversity needs to be taken far more seriously (Johnston *et al.*, 2012). The process of rebalancing the domains of economy, society, and environment can be advanced through a more inclusive approach to water governance which prioritizes and protects cultural diversity. But how can this be achieved when one culture, that of Western-derived rational materialism, considers itself to be the epitome of social and economic evolution? There is no room for other ontologies if a powerful consensus insists there are no viable alternatives.

Fortunately, the prevailing rational-materialist water ontology is experiencing some cracks. The unfolding climate crisis is undermining the moral legitimacy of neoliberal economic policies and motivating a new interest in Indigenous approaches to natural resources management, including water. At the same time, the cultural rights of Indigenous Peoples have been clarified and acknowledged through the 2007 UN Declaration on the Rights of Indigenous Peoples. Though still in the realm of soft law and without legal teeth, this combination of appreciation for and rights of Indigenous cultures can be seen reflected, to some extent, in water planning and (to a lesser extent) water policies. The following section discusses progress in creating space for imagining other water ontologies through: (1) Appreciating Indigenous knowledge, (2) Respecting Indigenous cultural rights, and (3) Indigenous water planning.

6.3.1 Appreciating indigenous knowledge

Indigenous and traditional cultural values and practices have long attracted the attention of economic development experts concerned about avoiding obstacles to project objectives, and identifying culturally informed strategies for inducing behavioral change. A World Bank collection of papers on the role of culture in economic development (Rao & Walton, 2004) suggested that cultural values can bend to the logic of rational materialism provided that the right enabling conditions of economic policies, social services (health and education), and infrastructure (roads, electricity) are in place. The message was that Indigenous cultures will, and essentially *should*, make way for a neoliberal agenda of economic development. In this view we need to understand Indigenous cultures, but only in order to change them.

The rise of sustainability as an overarching aim of economic development in the 1980s and '90 s, and in particular the 1992 Rio Conference on Sustainable Development, introduced powerful new reasons to study Indigenous cultures. Traditional ecological knowledge (Berkes *et al.*, 2000; Huntington, 2000) and 'Indigenous Knowledge' (Berkes, 2012 pp. 3–5) became viewed as important not only for particular Indigenous communities, but more generally for the larger project of sustainable development. This broad category of traditional/indigenous knowledge includes very specific technical knowledge about managing particular natural resources, for example, Balinese rice farmers' knowledge about controlling insect pests through careful timing of flood irrigation, as well as institutional knowledge about establishing roles and responsibilities for managing traditional irrigation canals in Nepal (Ostrom, 1992). Not only was such knowledge useful for designing more effective rural development interventions, but also for improving natural resources management within even the most developed economies (e.g., Ostrom, 2005).

Categorizing Indigenous knowledge as useful and potentially relevant beyond the Indigenous context marked a paradigm shift in legitimizing Indigenous cultural traditions in the eyes of the outside world. In addition to the ongoing initiatives to protect Indigenous cultures through the instruments of international human rights law (Anaya, 2000), there was now an additional cohort of allies: International development agencies and environmental experts interested in best practices for environmental sustainability. Recognizing the link between Indigenous knowledge and sustainability, in 2004 the annual UNDP World Development Report featured cultural diversity under the title, *Cultural Liberty in Today's Diverse World* (UNDP, 2004).

But while traditional cultures were clearly being taken more seriously, the global world order was selective in which parts of culture would be incorporated into mainstream policies. The values and ethics underlying Traditional Ecological Knowledge (TEK) tended to be overlooked. Deborah McGregor (2004, p. 389) points out that traditional 'principles and values such as respect, coexistence, cooperation, honor, thanksgiving, reciprocity, balance and harmony, and recognition of interrelationships among all of Creation' are traditionally communicated through myths and stories. 'Ultimately, TEK is related to Indigenous rights. Indigenous Knowledge cannot be separated from the people. This means that to protect IK or TEK, the people themselves and their ways of life must be protected' (Ibid, p. 399). It also implies that to fully understand TEK, one would need to study the cultural context of TEK, including the cultural values and environmental ethics that give rise to particular forms of TEK.

6.3.2 Respecting indigenous cultural rights

One of the defining features of Indigenous Peoples is that they are culturally distinct. They have their own way of doing things, and their own reasons for doing them, and their identity as a social group is tied to their shared cultural identity. Their right to culture is recognized in UN Resolutions and most specifically in the 2007 UN Declaration on the Rights of Indigenous Peoples (DRIP) which has become an indispensable tool of Indigenous activism (United Nations, 2008). Article 25 of the Declaration states that 'Indigenous peoples have the right to maintain and strengthen their distinctive spiritual relationship with their traditionally owned or otherwise occupied and used lands, territories, waters and coastal seas and other resources and to uphold their responsibilities to future generations in this regard.' Article 32 addresses the delicate issue of control over natural resources and makes three points:

(1) Indigenous peoples have the right to determine and develop priorities and strategies for the development or use of their lands or territories and other resources.
(2) States shall consult and cooperate in good faith with the indigenous peoples concerned through their own representative institutions in order to obtain their free and informed consent prior to the approval of any project affecting their lands or territories and other resources, particularly in connection with the development, utilization or exploitation of mineral, water or other resources.
(3) States shall provide effective mechanisms for just and fair redress for any such activities, and appropriate measures shall be taken to mitigate adverse environmental, economic, social, cultural or spiritual impact.

6.3.3 Whose laws? recognizing customary water rights

Under the Doctrine of Discovery proclaimed by the Vatican during the 15th and 16th centuries, the Portuguese and Spanish Crown adopted very explicit policies of ignoring the customary land and water rights of Indigenous Peoples. This Doctrine established an unfortunate legal precedent invoked by the US Supreme Court in 1823 and never formally repealed. Indeed, many contemporary national legal frameworks, including US law, still incorporate references to the Doctrine as a justification for state expansion into traditional indigenous territory (Miller, 2005:2–3). The moral legacy of this ethic is seen today in the general absence of state recognition of customary water rights (van Koppen, 2007).

An example of the practical impacts of not recognizing customary water rights is the case of the Pyramid Lake Paiute Tribe in the state of Nevada (USA). Traditionally the Indigenous community relied for their subsistence on the once plentiful fish from Pyramid Lake, a closed basin lake that has supported the tribe for the past 4,000 years. The lake is fed by the Truckee River which flows out of the Sierra Nevada mountains. Shortly after 1900, the US government constructed a dam on this river, upstream of the lake, to divert water into an irrigation canal destined for Euro-American farmers some distance away. The dam diverted about half the flow of the river, launching a cascade of ecological interactions: the lake level dropped, salinity increased, and fish could no longer reach the upper river to spawn. The result was the near decimation of the Indian tribe, bereft of its primary source of subsistence (Wilkinson, 2010). Court rulings during the 1970s to 1990s found that the US government had acted wrongly in ignoring the customary water rights of the tribe, but ruled that since so much time had elapsed, there was no practical way to compensate the tribe for their degraded lake (Wilkinson, 2010:220–2).

Similar cases abound where Indigenous communities relied on water supplies that were later wrested from their control. In the Northwest United States, the coastal Indian tribes, who relied traditionally on salmon, have seen fish populations collapse from water diversions and hydroelectric dams (Fisher, 2012). In the Andes, Indigenous farmers are engaged in a similar struggle: the irrigation canals first constructed by their Incan ancestors are being expropriated by state agricultural and irrigation agencies (Boelens *et al.*, 2007).

The practical outcome of realizing the cultural rights articulated in the UN Declaration hinges on the capacity of Indigenous communities to act collectively to claim these rights, and also depends on national laws and policies. Both these levels – internal cohesion and national politics – are complex and dynamic. Indigenous communities, like all communities, are composed of different types of people, reflecting a range of personalities and power relationships (Douglas, 2004). Representation of Indigenous Peoples' interests is further complicated by political institutions and local governance systems that are legacies of colonialism (Walker *et al.*, 2013). A further challenge within the water sector is the underrepresentation of Indigenous scientists and scholars who have few role models or peer guidance as they try to apply their traditional water values to their chosen scientific fields (Chief, 2018).

6.3.4 Reclaiming indigenous water planning

How can Indigenous communities participate meaningfully in water planning conducted by agents of the mainstream society, when there are fundamentally different ideas about water itself? Planning needs to undergo a process of 'unlearning' its Euro-American (colonial) assumptions about the nature of water (Porter, 2010). 'There is an ethic or duty of care to the land, its resource, and environments, particularly those located within the traditional territories of the group….Indigenous planning is place-based and implies a long and close association…of the specific environment and what it can sustain' (Jojola, 2013).

One promising approach for integrating Indigenous water values into the planning process is the systematic formulation of Indigenous water values into a formal water statement or declaration. The 2003 Indigenous Peoples' Kyoto Declaration on Water (discussed above) was a declaration at the global scale. In Australia, the Echuca Declaration was endorsed in 2007 by representatives of 31 Aboriginal communities, to formalize the concept of 'cultural flows' (Taylor *et al.*, 2016). A similar approach of formulating place-based local statements about water offers the opportunity to define Indigenous water values in Indigenous terms, while also delineating specific management goals, such as environmental and cultural flows. Through such statements and declarations Indigenous communities can control their own narrative about water.

In British Columbia, the Simpcw First Nation Water Declaration shares many concepts of the earlier Kyoto Declaration, but tailors the provisions to local circumstances. In particular, the Simpcw Declaration builds on the UN Declaration on the Rights of Indigenous Peoples, to specify how the concept of free, prior, and informed consent should be applied to negotiations involving the Simpcw Nation (Reading *et al.*, 2011, pp 101–108). For example, the declaration specifies that the consultative process must include the involvement of spiritual authorities as well as secular (governmental) authorities. Also, the Declaration makes explicit that if the consultative process does not result in agreement to whatever development project is being proposed (e.g., a dam or mine), that there must be 'Respect for the right to say no' (Ibid, p. 106).

In the Kimberley region of Western Australia, Indigenous people (referred to as Traditional Owners) agreed on common principles to protect the Fitzroy River catchment from mining and oil exploration. The Fitzroy River Declaration was proclaimed in November 2016, to serve as a basis for any future management decisions about the river and its catchment (Lim *et al.*, 2017). The Declaration states that, 'The Fitzroy River is a living ancestral being and has a right to life. It must be protected for current and future generations, and managed jointly by the Traditional Owners of the river.' The Owners agree to work together to (among other things):

- Reach a joint position on fracking in the Fitzroy catchment;
- Create a buffer zone for no mining, oil, gas, irrigation and dams in the Fitzroy catchment;
- Develop and agree a Management Plan for the entire Fitzroy Catchment, based on traditional and environmental values;
- Develop a Fitzroy River Management Body for the Fitzroy Catchment, founded on cultural governance;
- Complement these with a joint Indigenous Protected Area over the Fitzroy River.

6.4 ETHICS AS A BRIDGE ACROSS DIVERSE WATER CULTURES

Strategies of 'water diplomacy' (Islam & Susskind, 2013), which presume a negotiation process among roughly equal partners, are less useful when one side holds the power to force a settlement. And even if the powerful party has the best of intentions, the cultural values and ethics implicit in the language and categories (e.g., the concept of water 'resources') skews the frame towards the worldview of the politically dominant party. In the case of the Whanganui River in New Zealand, Maori attorneys argued successfully that the river should be accorded legal standing as a person (O'Donnell & Talbot-Jones, 2018). This legal fiction has a basis in both the Western legal framework and in Maori customary law; both legal systems recognize that 'personhood' can be applied to things other than human people. Under the agreement, two guardians, one from the Crown and one from a Whanganui River Indigenous community (*iwi*) have been assigned the role of protecting the river. A whole river strategy, in collaboration with *iwi*, local government, and commercial and recreational users is being formulated. The Whanganui River is the first river in the world to be recognized as a legal entity, and to my knowledge, still the only case that has withstood challenges at the highest level of national legal authority. (Personhood status was conferred on both the Yamuna and Ganga Rivers by state courts in India but was later overturned by the Indian High Court.)

6.4.1 Legal pluralism

The legal term for the art of seeing the same things from two very different perspectives is 'legal pluralism.' If we are serious about embracing cultural diversity then embracing legal pluralism offers a way of operationalizing that diversity. The Western legal system and the water ethics that underlie the laws and – the word fits: customs – of that legal system are not going to disappear very soon. In order to retain their cultural integrity, Indigenous groups will need to find ways of fitting their customary water behavior and ethics, within that Western reality. Legal pluralism offers a ready-made label that can help legitimize indigenous water ethics to non-Indigenous stakeholders (Meinzen-Dick & Nkonya, 2007). Granting legal personhood to rivers is a way of using Western legal concepts to Indigenous advantage and is promising in this regard, even though the Whanganui remains unique (Photo 2).

Photo 2 Are dams ethical? Aerial view of Cochiti Dam and reservoir, New Mexico, USA.

6.4.2 Relational ethics

The term, 'All our relations' which is used by Native North American as well as other Indigenous Peoples, has a real significance. The term connotes a merging of human kinship relations and kinship with the natural world, acknowledging ethical responsibilities implicit in those relationships. In her study of cooperative river planning for the Hurunui River in New Zealand, Thomas (2015, p. 984) notes that for the Maori, 'ethics of respect…extend to both human and nonhuman, where both care and use are equally central to livelihoods and identity.' Maori acknowledge both economic concerns of people and concerns for the river, attempting 'to reach a balance between responsibility for the Hurunui River, and demands to produce economic growth' (Ibid, p. 984). A key value is placed on 'a more reciprocal relationship with the Hurunui River and [Maori's] rights, interests, uses and cultural values' (Ibid, p. 984). A practical outgrowth of this reciprocity value was to include in the river plan the restoration of the lagoons and delta at the mouth of the river as a way to compensate the river for increased upstream water diversions which the plan was also recommending. Thomas recounts that non-Indigenous members of the river planning group gained a new respect for the stewardship ethics of the Maori members. Thomas further observes that when non-Indigenous members invoke similar ethics about stewardship and giving back to the river, they lacked the same legitimacy. Citing McGregor (2004, p. 602) this differential legitimacy can be explained as due to 'the lack of an established alternative vocabulary' or in the words of Thomas (2015, p. 985), 'The absence of a coherent non-Maori narrative based on spiritual connections that would challenge the limitations of Eurocentric binaries.'

6.5 CONCLUSIONS: WATER ETHICS AND CO-MANAGEMENT

The Hurunui River planning committee discussed above is an example of cross-cultural stakeholder co-management built on a foundation of shared values and ethics about the river itself (relational ethics) and ethics about the appropriate uses of water (e.g., for irrigation). Through the planning committee (This refers to the Hurunui Waiau Zone Committee (HWZC) which is part of a regional initiative, the Canterbury Water Management Strategy (Thomas, 2015, p. 979).) both Maori and non-Maori representatives shared their value perspectives about the river. The common language for describing their value priorities was, essentially, the language of ethics. The Maori representatives in this case served as teachers to the non-Maori, explaining how the Maori regarded the river, why they felt a kinship with the river, and a responsibility to act on behalf of the river, etc. When the Maori explained their ethics, the non-Maori could see a correspondence with their own values as well, but that breakthrough required some education. More precisely, it required the non-Maori to 'de-learn' their own colonial assumptions about the nature of rivers and how rivers should be treated. Only after this informal education process could constructive co-management of the river, or even the constructive formulation of a river plan, proceed.

1. What are the implications for bridging diverse water cultures, and establishing effective co-management strategies for diverse stakeholders? A common language of water ethics needs to be established as a prerequisite to anything else, but to even talk in ethical terms about water requires an awareness about values relevant to water. In my 2013 book, *Water Ethics: A Values Approach to Solving the Water Crisis* (Groenfeldt, 2013) (A second revised edition of this book was published in 2019.) I distinguish five categories of water ethics: Environmental, Economic, Social, Cultural, and Governance. Each category contains a small world of values about water; for example, social values include social and gender equity, safety (clean water), the right to affordable water, the right to enjoy healthy rivers, etc. Governance values include participatory planning, transparency and accountability, and democratic decision processes, as well as values about professional competence, responsible use of technologies, etc. It is probably not necessary for members of a watershed planning committee to learn every aspect of water ethics, but they do need to at least recognize that the category of ethics can be applied to water.

The most important dimension of Indigenous water knowledge – the ethical norm of being in relationship to the water and to all the people, plants, and animals that also depend on water – has been largely overlooked. By carving out a category of 'water ethics' in our collective ontologies, we will have a place for processing the wisdom that Indigenous Peoples, both past and present, can teach us. Through learning about Indigenous understandings of water, we will inevitably gain an appreciation and a sense of respect for their way of seeing the world. Once we start to operationalize our 'duty to learn' about Indigenous water ontologies, we will find useful applications of the knowledge we gain. The challenges of our changing climate and the crisis of sustaining our water resources will provide many practical opportunities.

REFERENCES

Anaya J. (2000). Indigenous peoples in international law. Oxford University Press, New York.

Arthington A. H., Bhaduri A., Bunn S. E., Jackson S. E., Tharme R. E., Tickner D., Young B., Acreman M., Baker N., Capon S., Horne A. C., Kendy E., McClain M. E., Poff N. L., Richter B. D. and Ward S. 2018. The Brisbane declaration and global action agenda on environmental flows (2018). *Frontiers in Environmental Science*, **6**(45). www.frontiersin.org/articles/10.3389/fenvs.2018.00045/full

Basso K. H. (1996). Wisdom Sits in Places: Landscape and Language among the Western Apache. UNM Press, Albuquerque, New Mexico.

Berkes F. (2012). Sacred Ecology, 3rd edn. Routledge, New York.

Berkes F., Colding J. and Folke C. (2000). Rediscovery of traditional ecological knowledge as adaptive management. *Ecological Applications*, **10**(5), 1251–1262.

Boelens R., Bustamante R. and de Vos H. (2007). Legal pluralism and the politics of inclusion: recognition and contestation of local water rights in the Andes. In: Community-based Water Law and Water Resource Management Reform, B. van Koppen, M. Giordano and J. Butterworth (eds.), CAB International, London.

Boelens R., Chibba M. and Nakashima D. (eds) (2006). Water and Indigenous Peoples. UNESCO, Paris. http://unesdoc.unesco.org/images/0014/001453/145353e.pdf

Chen C., MacLeod J. and Neimanis A. (eds.) (2013). Thinking with Water. McGill- Queen's Press, Montreal and Kingston.

Chief K. (2018). Emerging voices of tribal perspectives in water resources. *Journal of Contemporary Water Research & Education*, **163**(1), 1–5.

Cruikshank J. (2012). Are glaciers 'good to think with'? Recognising indigenous environmental knowledge. *Anthropological Forum*, **22**(3): 239–250.

Douglas M. (2004). Traditional culture: let's hear no more about it. In: Culture and Public Action, V. Rao and M. Walton (eds), Stanford University Press, Stanford, CA.

Finch L. S. G. (2012). The duty to learn: Taking account of indigenous legal orders in practice. Paper presented at Indigenous Legal Orders and the Common Law, Continuing Legal Education Society of British Columbia, November 15, Vancouver, BC.

Fisher A. (2012). Spirit of the salmon: native religion, rights, and resource use in the Columbia river basin. In: Indigenous Knowledge and the Environment in Africa and North America, D. Gordon and S. Krech, III (eds), Ohio University Press, Athens, OH.

Groenfeldt D. (2013). Water Ethics: A Values Approach to Solving the Water Crisis. Routledge, New York.

Groenfeldt D. (2016). Cultural water wars: power and hegemony in the semiotics of water. In: The Politics of Fresh Water: Access, Conflict and Identity, C. M. Ashcraft and T. Mayer (eds), Routledge, London, pp. 143–156.

Groenfeldt D. (2019) Water Ethics: A Values Approach to Solving the Water Crisis, 2nd edition. Routledge, NY.

Hayman E., Wedge M. and James C. (2017a). A deep chart (the aqua-face of deep mapping): collaborative water research with Carcross/Tagish first nation, Shaana- khéeni headwaters, Yukon territory/British Columbia, Canada. *International Journal of Humanities and Arts Computing*, **11**(1), 86–108.

Hayman E., James C., Wedge M. and Katzeek D. (2017b). Marrying the water: A Tlingit and Tagish approach towards an ethical relationship with water. IN: Global Water Ethics: Towards a Global Ethics Charter,R. Ziegler, and D. Groenfeldt (eds), Routledge, London.

HLPW. 2017. Bellagio principles on valuing water. High Level Panel on Water, May (2017). https://sustainabledevelopment.un.org/content/documents/15591Bellagio_principles_on_valuing_water_final_version_in_word.pdf

Howarth W. (2018). Going with the flow: integrated water resources management, the EU water framework directive and ecological flows. *Legal Studies*, **38**, 298–319.

Huntington H. P. (2000). Using traditional ecological knowledge in science: methods and applications. *Ecological Applications*, **10**, 1270–1274.

International Food Policy Research Institute (IFPRI); and VEOLIA. (2015). The murky future of global water quality, White Paper from IFPRI and Veolia Water North America. http://ebrary.ifpri.org/cdm/ref/collection/p15738coll2/id/129349

Iorns Magallanes C. J. (2018). From rights to responsibilities using legal personhood and guardianship for rivers. In: Responsability: Law and Governance for Living Well with the Earth, B. Martin, L. Te Aho and M. Humphries-Kil (eds), Routledge, London & New York, pp. 216–239.

Islam S. and Susskind L. E. (2013). Water Diplomacy: A Negotiated Approach to Managing Complex Water Networks. RFF Press, New York.

Jackson S. and Barber M. (2016). Historical and contemporary waterscapes of North Australia: indigenous attitudes to dams and water diversions. *Water History*, **8**, 385–404.

Johnston B. R., Hiwasaki L., Klaver I. J., Ramos Castillo A. and Strang V. (eds) (2012). Water, Cultural Diversity, and Global Environmental Change: Emerging Trends, Sustainable Futures?. UNESCO, Paris and Springer, Dordrecht, Germany. http://unesdoc.unesco.org/images/0021/002151/215119E.pdf

Jojola T. (2013). Indigenous planning: towards a seven generations model. In: Reclaiming indigenous planning, R. Walker, T. Jolola and D. Natcher (eds), McGill-Queen's University Press, pp. 457–472.

Lim M., Poelina A. and Bagnall D. (2017). Can the Fitzroy river declaration ensure the realisation of the first laws of the river and secure sustainable and equitable futures for the West Kimberley? *Australian Environment Review*, **32**(1), 18–24.

McGregor D. (2004). Coming full circle: indigenous knowledge, environment, and our future. *American Indian Quarterly*, **28**, 385–410.

Meinzen-Dick R. S. and Nkonya L. K. (2007). Understanding legal pluralism in water and land rights: lessons from Africa and Asia. In: Community-based Water Law and Water Resource Management Reform in Developing Countries, B. van Koppen, M. Giordano and J. Butterworth (eds), CAB International, London, pp. 12–27.

Miller R. (2005). The doctrine of discovery in American Indian law. *Idaho Law Review*, **42**, 1–122.

O'Donnell E. L. and Talbot-Jones J. (2018). Creating legal rights for rivers: lessons from Australia, New Zealand, and India. *Ecology and Society*, **23**(1), 7. https://doi.org/10.5751/ES-09854-230107

Ostrom E. (1992). Crafting Institutions for Self-Governing Irrigation Systems. ICS Press, San Francisco.

Ostrom E. (2005). Understanding Institutional Diversity. Princeton University Press.

Porter L. (2010). Unlearning the Colonial Cultures of Planning. Ashgate, Burlington, Vermont.

Rao V. and Walton M. (eds) 2004. Culture and Public Action. Stanford University Press, Stanford, CA.

Reading J., Perron D., Marsden N., Edgar R., Saravana-Bawan B. and Baba L. (2011). Crisis on tap: Seeking solutions for safe water for Indigenous peoples. Aboriginal Policy Research Consortium International (APRCi). Paper 211. http://ir.lib.uwo.ca/aprci/211

Rokeach M. (2000). Understanding Human Values, 2nd edition. Simon and Schuster, New York.

Suzuki D. and Knudtson P. (1992). Wisdom of the Elders: Honoring Sacred Native Visions of Nature. Bantam Books, New York.

Taylor K. S., Moggridge B. J. and Poelina A. (2016). Australian indigenous water policy and the impacts of the ever-changing political cycle. *Australasian Journal of Water Resources*, **20**(2), 132–147.

Thomas C. A. (2015). Indigenous more-than-humanisms: relational ethics with the Hurunui river in Aotearoa New Zealand. *Social & Cultural Geography*, **16**, 974–990.

UNDP. (2004). Human Development Report: Cultural Liberty in Today's Diverse World. United Nations Development Program, New York. http://hdr.undp.org/sites/default/files/reports/265/hdr_2004_complete.pdf

United Nations. (2008). United Nations Declaration on the Rights of Indigenous Peoples. United Nations, New York. www.un.org/esa/socdev/unpfii/documents/DRIPS_en.pdf

van Koppen B. (2007). Dispossession at the interface of community-based water law and permit systems. In: Community-based Water Law and Water Resource Management Reform, B. van Koppen, M. Giordano and J. Butterworth (eds), CAB International, London.

Veilleux J. (2017). Aquatic system connectivity and the case of standing rock. Water Resources Impact, March, 32–34.

Walker R., Jolola T. and Natcher D. (eds) (2013). Reclaiming Indigenous Planning. McGill-Queen's University Press, Montreal and Kingston.

Wilkinson C. F. (2010). Indian water rights in conflict with state water rights: the case of the pyramid lake paiute tribe in Nevada, US. In: Out of the Mainstream: Water Rights, Politics and Identity, R. Boelens, D. Getches and A. Guerva-Gil (eds), Earthscan, London.

Wolf A. (2017). The Spirit of Dialogue: Lessons from Faith Traditions in Transforming Conflict. Island Press.

Yates J. S., Harris L. M. and Wilson N. J. (2017). Multiple ontologies of water: politics, conflict and implications for governance. *Environment and Planning D: Society and Space*, **35**(5), 797–815.

Chapter 7

Foggara water which left an imprint in the culture of oases of the southwest region of Algeria

Ansari Taha

ABSTRACT

In 2006, the UN Secretary-General Kofi Annan said in his message during the World Water Day in 2006, with the theme 'water and culture', that water is not only essential for life; it is also a wide-ranging cultural presence – an inspiration for artists, a focus of scientific research, and an indispensable element in the religious rituals of many traditions and faiths. All this shows the importance of water in different cultures and shows that water is a Social and Cultural Space. In arid landscapes like the south-west region of Algeria, water is at the heart of human existence and the development of this region. The people have managed this precious resource and have developed many cultural practices around water. The people gave their water management system of underground channels the name 'foggara'. The aim of this study is to understand the purpose of foggaras and to examine how foggaras – as an essential system for life in an arid area – play an important, conceptual role in local cultural orientations. The people of this region have built an environment that respects the laws of sustainable development, that is to say the oasis. An oasis is composed of a source of water (foggaras), a village (ksar), and a palm-grove. The foggara water in an oasis has a strong cultural dimension. Without understanding and considering the cultural aspects of our water problems, no sustainable solution can be found.

Keywords: Foggara, oases, culture, groundwater, arid region

Editor: Xiao Yun Zheng
doi: 10.2166/9781789062045_0101

7.1 INTRODUCTION

The Algerian Sahara is the third-largest and hottest desert in the world. Water is scarce in this very arid region, with annual precipitation rates of less than 100 mm and as little as 50 mm in some places. The people of these regions have learnt how to adapt to the hostile climate and understand how to tame their inhospitable environment. Water is fundamental to human existence and human beings cannot survive for long without it. Since surface water is rare, the inhabitants of these regions have begun to search for groundwater. One of the means used for exploiting groundwater in the desert is the system of underground channels known as foggaras. The foggaras are located in the southwest of Algeria in the province of Adrar. The topography and hydrogeology of Adrarare propitious for creating foggaras because there is a natural land slope and groundwater is not very deep. Adrar's is divided into three regions, Touat, Gourrara and Tidikelt, where foggaras are located.

The people of the Sahara have drilled almost 280,000 wells and have dug about 3,200,000 km of underground canals to make connection between wells. They have built huge irrigation networks inside palm-groves so as to use this water. In addition, there is a close relationship between foggara's water and culture and the customs and traditions practiced in the region of Adrar. All through history, the communities of Touat, Gourrara, and Tidikelt has led the development of the culture of cooperation and an improved culture of management of this water to avoid any social conflict and problems inside the ksar (village). The culture of cooperation has embodied all activity related to the foggaras, so that all inhabitants of the ksar are mobilized when there is a problem affecting the foggaras. Concerning the culture of the management of the water of foggara, the man in charge of an oasis uses his knowledge to promulgate laws depending on the Islamic laws and his know-how. Foggara is one of the traditional systems in the world that have created civilization with a rich culture, and this culture is still alive today.

7.2 ADRAR, THE LAND OF FOGGARAS

7.2.1 Geographic location

The province (*wilaya* in Arabic) of Adrar is located in the central part of the Algerian Sahara and covers a total area of 427,968 km^2 (Figure 7.1). It has a population of about 432,000 inhabitants. Adrar is bordered in the:

- Northwest by the province of Bechar
- East by the provinces of Ghardaia and Tamanrasset
- West by the province of Tindouf
- Southwest by the Republic of Mauritania
- South by the Republic of Mali.

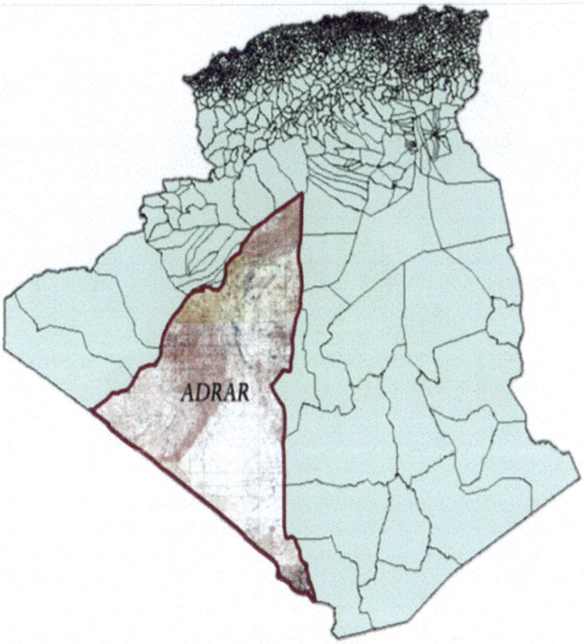

Figure 7.1 Location of Adrar province in Algeria. (Extract from GIS in National Agency of hydraulics Resources [ANRH in French].)

Adrar is subdivided into four regions (Figure 7.2):

- Gourara
- Taouat
- Tidikelt
- Tanezrouft.

It is administratively divided into:

- 11 districts (*daïras* in French)
- 28 municipalities (*communes* in French).

7.2.2 Hydrology

It rarely rains in the Adrar. The few streams identified have intermittent flows that can cause sudden and violent floods.

7.2.3 Hydrogeology of Adrar

The groundwater used in most of the province of Adrar is drawn from the Continental Intercalary aquifer, which is part of the North Western Sahara Aquifer System (NWSAS). The NWSAS (in French, the Système Aquifère du

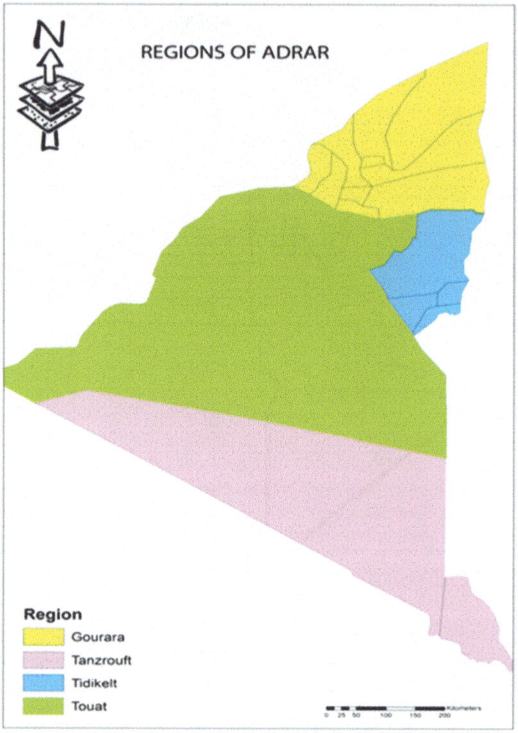

Figure 7.2 Regions of Adrar. (Extract from GIS in National Agency of hydraulics Resources [ANRH in French].)

Sahara Septentrional) covers a total area of over 1,000,000 km^2: 700,000 km^2 in Algeria, 80,000 km^2 in Tunisia, and 250,000 km^2 in Libya. It contains sedimentary deposits which, from bottom to top, have two main levels of aquifers: the Continental Intercalary and the Terminal Complex (Observatoire du Sahara et du Sahel, 2003). The thickness of the Continental Intercalary in Adrar varies from north to south and from east to west. It oscillates between 150 and 620 m. The static level of the aquifer varies from 6 to 30 m. In addition, this aquifer is exploited by boreholes, wells, and the foggaras.

7.3 DESCRIPTION OF FOGGARA

The foggara is an underground channel that drains the water of the Continental Intercalary and conveys it to the palm grove. A series of well-like vertical shafts of variable depths connect the foggara to the surface (Figures 7.3 and 7.4). The foggara ends in a comb-like structure called a kasria, which distributes the water in to a network of ditches (Figure 7.5).

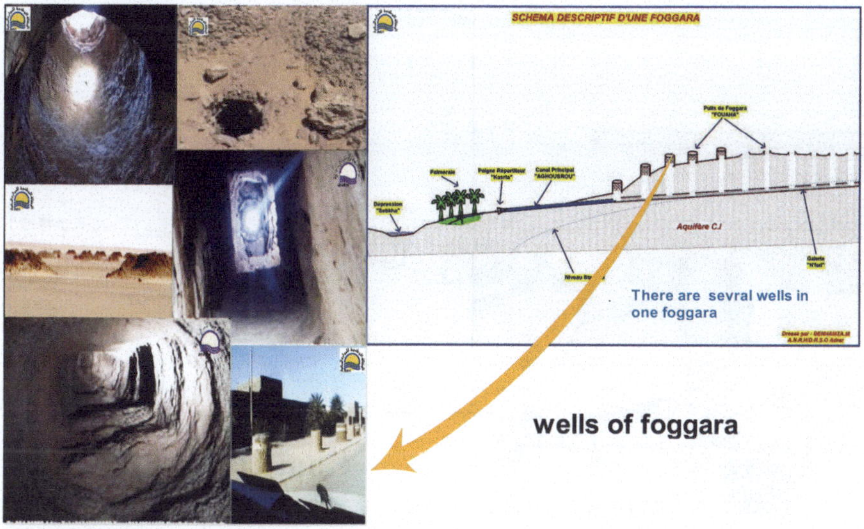

Figure 7.3 Wells of a foggara.

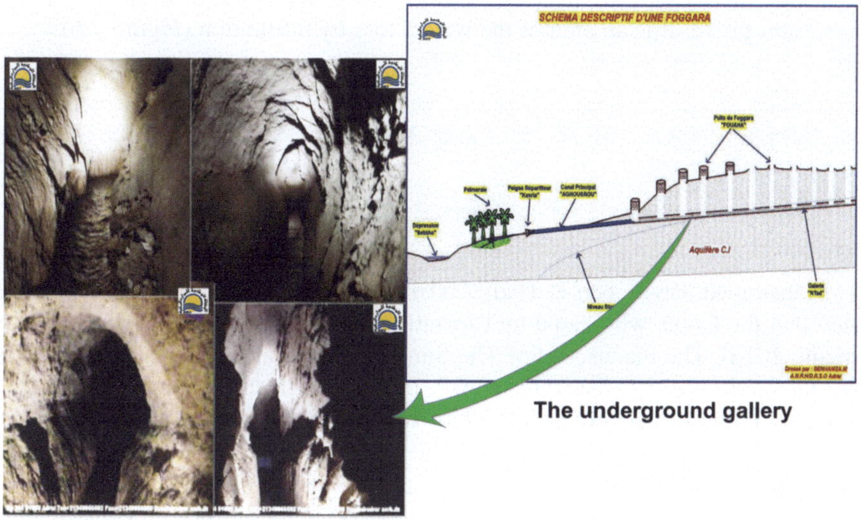

Figure 7.4 Underground channel.

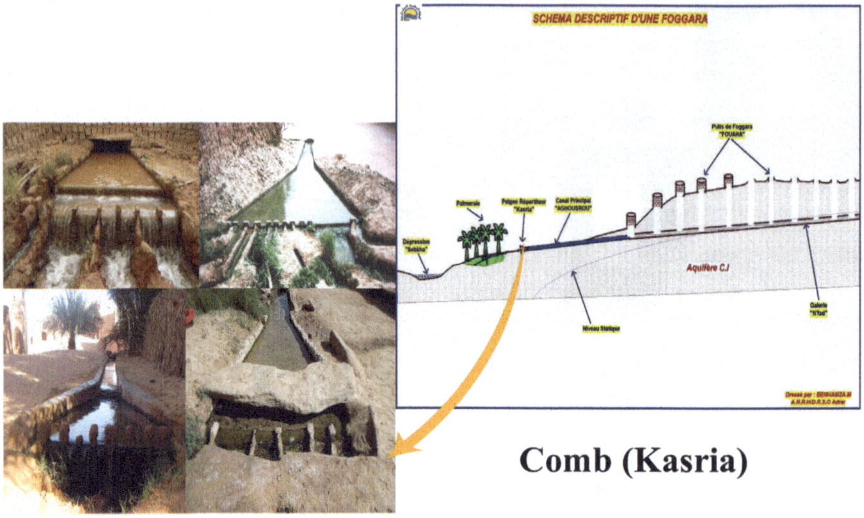

Figure 7.5 Comb (kasria).

7.4 OPERATION OF THE FOGGARA

The foggara is composed of two distinct parts:

- An active part located upstream situated in the aquifer. It is the most important part of foggara.
- The inactive part located downstream, which allows the flow of water to the palm-grove. It is an area of the waters loss by infiltration (Figure 7.6).

7.5 WHAT WAS WRITTEN ABOUT FOGGARAS IN MANUSCRIPTS

7.5.1 Manuscript 1: 'The Simple Speech in the News of Tamentit'

Mr. Mohammed Tayeb ben el Hadj Abderrahime, the author of this manuscript, stated that the Copts who came to Tamantit were the first people to dig foggaras (Ismaili, 2018). The manuscript of The Simple Speech in the News of Tamentit is considered as one of the rare works that highlighted areas of the social, economic, cultural, and urban history of the oasis of Tamantit. The manuscript revealed valuable information on the history of the settlement of tribes, the reconstruction of the place, and how palaces were constructed, and their development during an important historical period in the history of Touat region.

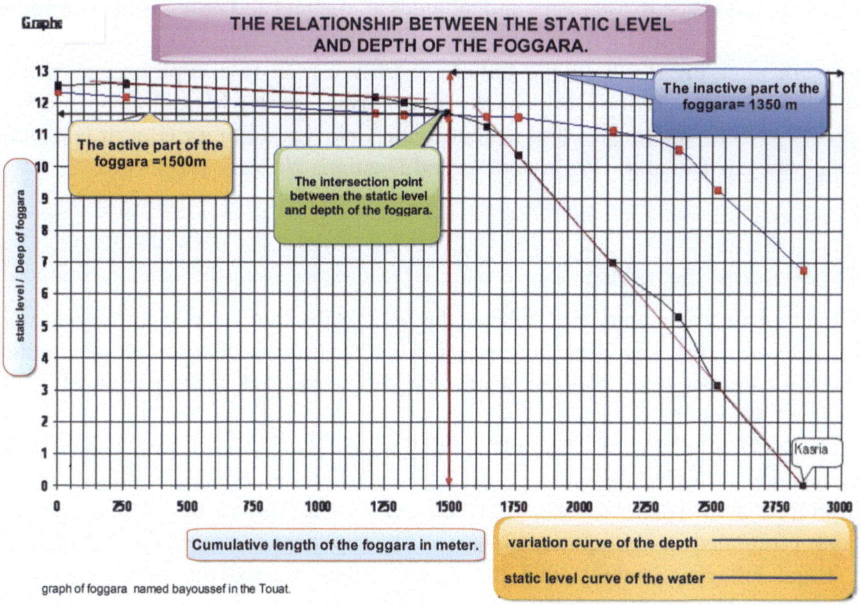

Figure 7.6 Relationship between static level and depth of foggara (foggara Bayoucef).

The author of the manuscript was Muhammad al-Tayyib ibn al-Hajj Abd al-Rahim al-Tawati al-Qurashi al-Tintati, known as Ibn Baba Haida from the Touat jurists in the ninth century. He was born in Safar 1148 AH in Tamantit (Tamantit is an oasis in Adrar province). His origin is from the family of the Ouled Yahya from the Minyara tribe in Morocco, which belongs to the Quraysh tribe. Al-Tayyib Abdel-Rahim lived in Tamantit (Mohammed, 2016).

7.5.2 Manuscript 2: 'The Book of the Questionnaire on what Happened to Touat of Matters and Issues'

The author of this manuscript declared that the Jewish people were the first to dig foggaras in Touat region. 'The Book of the Questionnaire on what Happened to Touat of Matters and Issues' was considered among the most important and largest manuscripts that brought together a large amount of new issues and jurisprudential decisions (fatwas) that occurred in Touat. And the senior scholars and jurists from the region, such as Sheikh Sidi Abd al-Rahman bin Baumer 1189 AH, Sheikh Zajlawi 1212 AH, and Sheikh Sidi Hajj al-Balbali 1244 AH and his son Sheikh Hajj Abdel Aziz 1261 AH and many others. so the manuscript consists of 700 pages and includes all matters and issues that occurred in the Touat region in southern Algeria throughout the ages. Sidi Abdel Aziz complete the work after the death of his father. There are now two copies of it, one in Balbalin's library (Belbali is a name of the tribe) at the Malouka (an oasis in Adrar province), and the second is kept in the Bakriyin's library (Bekrii is a name of the tribe) in Tamantit (Professor Ahmed Djaafari, 2009).

7.5.3 Manuscript 3: 'The Precious Pens in the News of the Maghreb after Islam'

The author of this manuscript believe that the Arabs first created the foggaras in Touat region. This was after their arrival from the Arabian Peninsula. The author argues for this suggestion on the basis that the word used in the region of Touat to name this system is Arabic.

The local scholars, for example, Mohammed ben El Bekri, believed that the Barmakides brought the concept of foggara from the Persian civilisation and introduced it in Touat. Mohammed ben Omar ben Ahmed ben Mohammed El Habib postulated in his manuscript titled 'The report of narrators of the creativity of Touat villages (ksours)' that the Zeneta tribes created the first foggaras in Touat; the scholar Ahmed ben Youcef Etililani in his manuscript titled 'The origin of Touat's dwellers' made the same suggestion. This hypothesis is also supported by Molay Abdel Allah Ismailli in his last book entitled 'IFLI (fogarra) Water allocation mechanism- genesis and evolution' (Ismaili, 2018). Molay Abdel Allah Ismailli supports this idea by pointing out that all nouns describing foggaras are the words of Zeneta dialect, and the names of tools used in palmgroves originate from the Zenata dialect.

7.6 INVENTORY OF FOGGARAS IN ADRAR

The National Agency of Hydraulics Resources made an inventory of foggaras in 1998. This covered all foggaras of the province of Adrar. The inventory was updated in 2011, 2012, 2013, 2014, 2015, 2016, and 2019 (ANRH, 2019). There are about 2310 foggaras in Adrar. The National Agency of Hydraulics Resources has also recorded locations of wells and kasrias through GPS for all foggaras of Touat and Tidikelt regions (Figures 7.7–7.11).

7.7 QUALITY OF THE WATER FROM FOGGARAS

The water from foggaras was used for both drinking and for irrigation in the past, but currently, the water of foggara is used only for irrigation purpose. The most important factors in determining the required quality of water in agriculture are: 1. pH; 2. Risk of Salinity (TDS); 3. Hazard of Sodium (Sodium Adsorption Ratio or SAR).

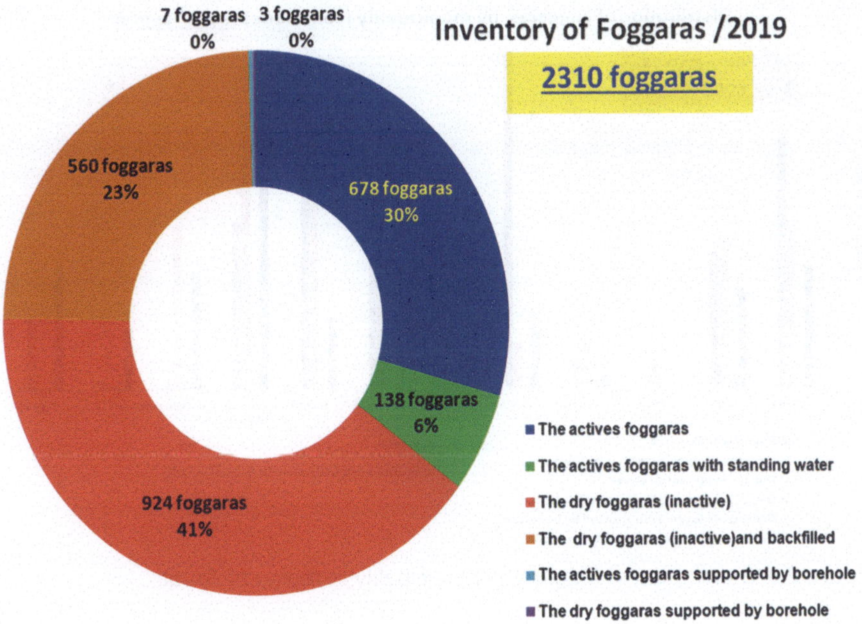

Figure 7.7 Inventory of foggaras in Adrar.

- The pH indicates that 82% of foggaras provide water in the ‘good’ category for agriculture.
- Total Dissolved Salinity (TDS) indicates that 64% of foggaras provide water that is in the ‘good’ category for agriculture, although 36% is used with caution.

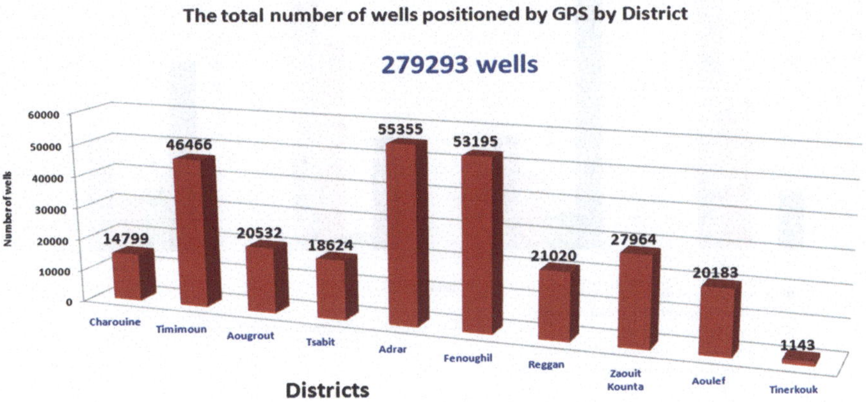

Figure 7.8 The total number of wells positioned by GPS by district.

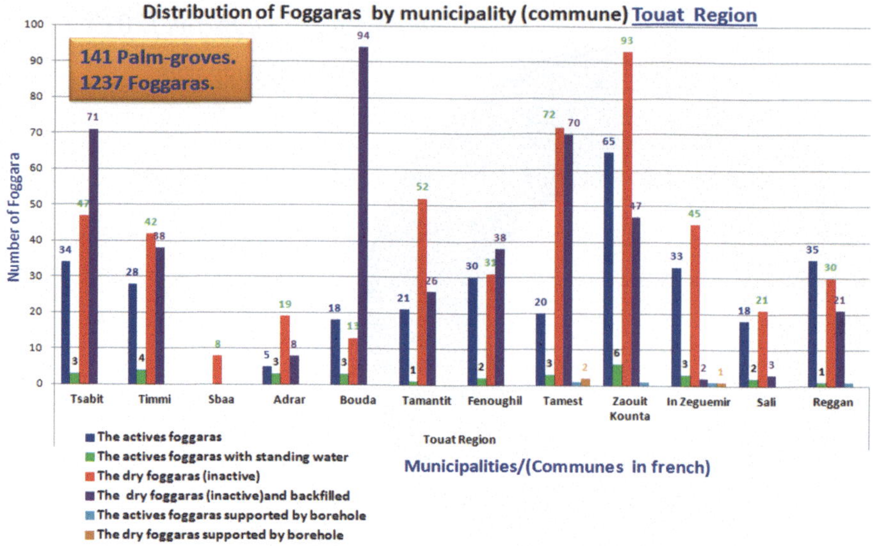

Figure 7.9 Distribution of foggara by municipality in region of Touat.

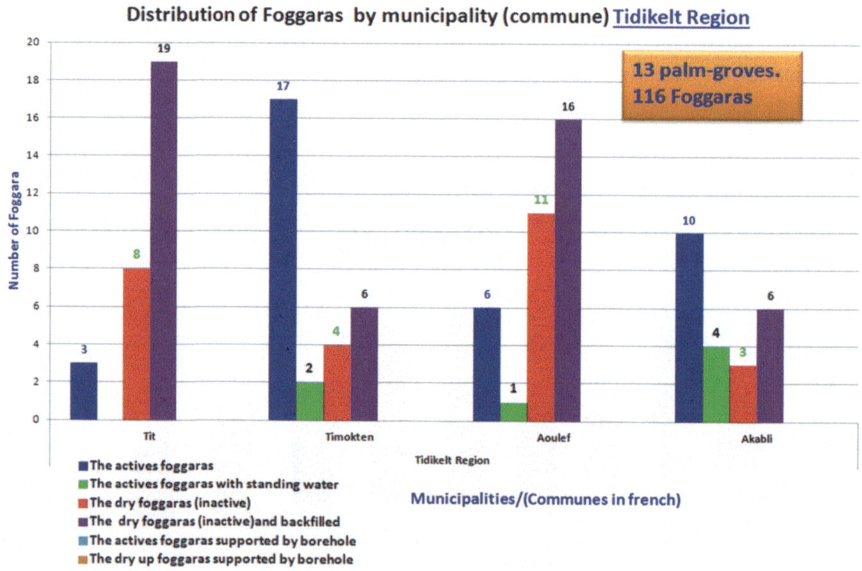

Figure 7.10 Distribution of foggara by municipality in region of Tidikelt.

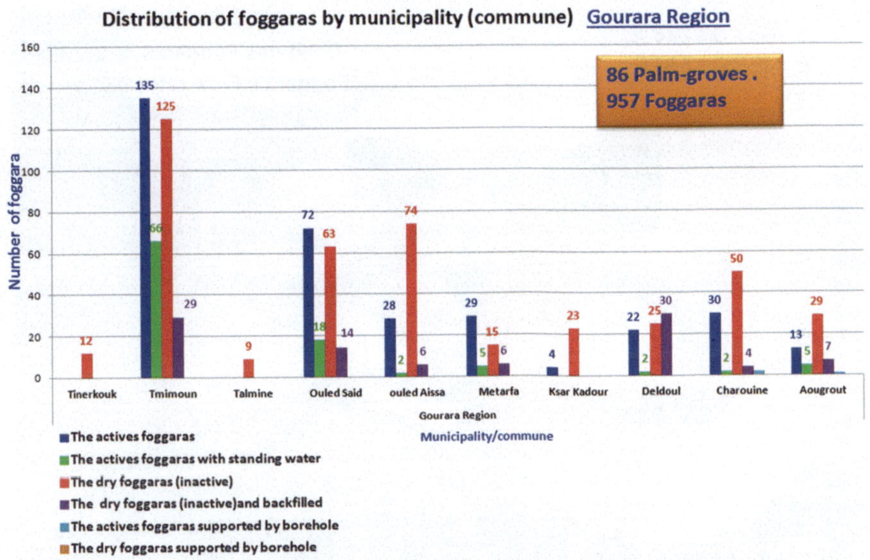

Figure 7.11 Distribution of foggaras by municipality in region of Gourara.

- 94.5% (688 foggaras) of foggaras provide low Na water (little danger); 5.21% (38 foggaras) provide medium Na water (problems on fine texture soils and sodium-sensitive plants, especially under low-leaching conditions. Soils should have good permeability); 0.14% (1 foggara) of foggaras provide very high Na water (Unsatisfactory) (Nasri, 2018).

7.8 MEASUREMENT OF FOGGARAS' FLOW RATES

In 2014, the National Agency of Hydraulics Resources updated the measurement of the flow rates of perennial foggaras by using a mini current meter (ANRH, 2012). The results are shown in Figure 7.12.

7.9 WATER OF FOGGARA IN CULTURE AND SOCIETY

A key message on water and culture was given on the occasion of World Water Day 2006 which promoted the understanding that Water and Culture are inseparable elements of human life: Culture should be regarded as the permanently evolving set of distinctive spiritual, material, intellectual and emotional features of society or a social group. It encompasses – in addition to art and literature – lifestyles, ways of living together, value systems, traditions and beliefs. The way water is used and valued constitutes an integral part of a society's cultural identity (World Health Organization, 2006).

Figure 7.12 Measurement of foggara flow-rates.

Water is at the heart of human existence and development. Nevertheless, how do people manage this precious resource in arid landscapes, such as in the Adrar region? What cultural practices have been developed by these communities of Touat, Tidikelt, and Gourrara for using and managing this water? Finally, what is the role of women in the formation of this culture around water?

The main objective here is to ascertain the relationship between water and culture. In this section, we have several specific objectives through which the main objective of the research would be achieved.

- First, identification of current social and cultural values and practices associated with water of foggaras.
- Second, analyzing the changes in these cultural and social systems and their significant influence on the system of foggaras.

7.9.1 Foggara in oases

There may be one or more foggara in a ksar (village). Each foggara has a name and each has an owner or owners. The owners may be the inhabitants of the ksar or another ksar. The owners are responsible for the maintenance, protection, and management of the foggara. The Algerian State requires the owners of foggaras to organize and form different associations. Each association has a president, financial secretary, and members who are foggara owners.

Figure 7.13 The maintenance and rehabilitation of a foggara.

7.9.2 Foggara and culture of cooperation (collective action)

Water is a scarce natural resource in this arid region. Foggaras are used to supply water for the irrigation of palm groves, and this important process requires everybody to have a sense of responsibility towards the foggaras. This sense of responsibility towards the vital resource has established the culture of cooperation between the inhabitants of ksar. If there is any damage or problems in a foggara, which cause a decrease in the quantity of water, the inhabitants of ksar, both men and women, will mobilize to repair the damage (see Figures 7.13, 7.14 and 7.15).

Figure 7.14 The cleaning of the foggara.

Figure 7.15 Foggaras and culture of team spirit.

There are meetings between owners of foggaras and administration during the repair of the foggara as shown in Figure 7.16.

The role of the woman is food preparation for the men who carry out physical work on the foggaras (Figure 7.17).

Figure 7.16 Collective action – a meeting during repair of a foggara.

Figure 7.17 Food for workers.

7.10 FOGGARA AND THE CULTURE OF WATER FOR PEACE AND CULTURE OF THE MANAGEMENT OF WATER

The establishment of a culture of peace requires the enactment of laws to protect and manage foggara water within oases. The first law is that the water of foggara is free for daily usage (see Figures 7.18–7.21). For agriculture, this water becomes a product to be sold, bought, and also bequeathed. This law is very useful in the culture of management and protection of the water of foggara. Each foggara has a very important document called in local dialect *Ezmmam el foggara*. It contains all the quantities of water of foggara allocated to each owner. Also, all quantities of the sale and the purchase were registered in it (Figures 7.22–25).

Figure 7.18 Water of foggara is free for daily usage.

Figure 7.19 Using foggara water for washing clothes.

Figure 7.20 Using foggara water for drinking.

7.11 FOGGARA CONSERVATION MEANS THE PROTECTION OF GENERATION CULTURE THROUGHOUT THE CENTURIES

The foggara in Algeria, the Qanat in Iran, Khettara in Morocco, and Falj in Oman are systems used for the acquisition and distribution of water based on horizontal

Figure 7.21 Using foggara water for washing utensils.

drainage galleries. However, although there are similarities in their collection system (galleries and shafts), these traditional techniques show differences in their water source and in the technique of water sharing (Remini *et al.*, 2010). In Algeria, the foggara represents the history of the communities of Touat, Tidikelt, and Gourrara. The real threat to this cultural heritage is the loss of the know-how and how to transmit knowledge and skills to future generations. Modern developments in Touat, Tidikelt, and Gourrara are promoting radical change in the culture of this society because drinking water supply networks have replaced

Figure 7.22 Traditional method of measurement of foggara's flow rate.

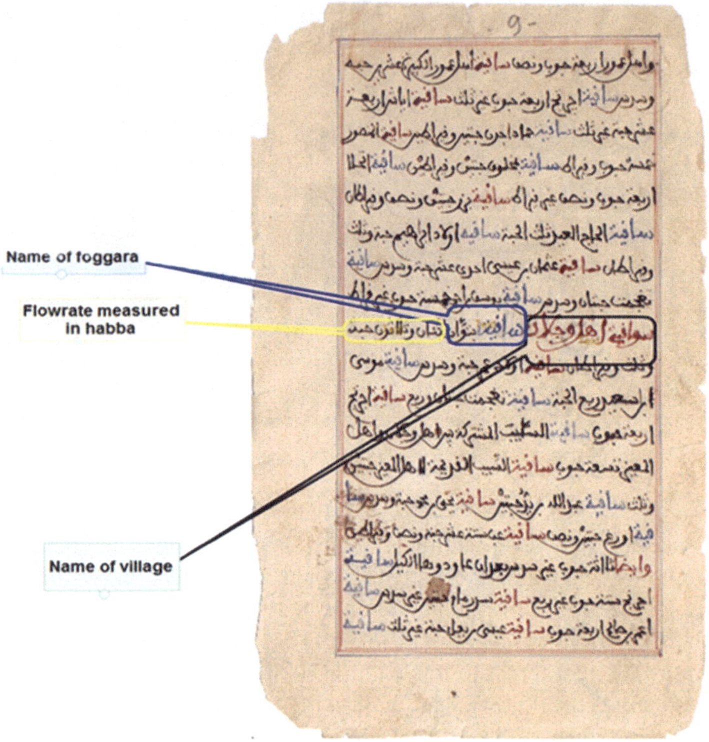

Figure 7.23 Traditional inventory and traditional measurment of foggara's flow rate.

Figure 7.24 Traditional tools to measurement of the foggara's flow rate.

Figure 7.25 Workers inside foggaras.

the foggara. Also, the transition from a seasonal agricultural economy based on the water supplied by the foggaras, to a more 'sustainable' pivot-irrigation agriculture, has made huge changes in the culture of oases.

At the request of Algeria, the knowledge and skills of the water measurers of the foggaras or water bailiffs of Touat and Tidikelt are inscribed in 2018 on the List of Intangible Cultural Heritage in Need of Urgent Safeguarding). Therefore, the element concerns the knowledge and skills of the water measurers of the

Figure 7.26 The art of measuring the foggara.

foggaras, or water bailiffs, of village communities of Touat, Gourrara, and Tidikelt. The water measurers are a very important persons in the oases and form the cornerstone of the culture of foggara. This is because the water measurers are involved in various operations, from calculating water shares to repairing distribution combs and conducting water in the channel (Figure 7.26).

7.12 CONCLUSION

The system of foggara is not only an original irrigation technique and art of measurement, it is also an 'ingenious and centuries-old' system which is part of the identity of the Saharan populations in general and those of the wilaya of Adrar in particular. The foggara is made up of 'knowledge, know-how and rituals enclosing many elements of the intangible heritage transmitted from generation to generation and going in the direction of ecological management of the resources of nature and rational exploitation of the water scarcity'. For this purpose, it is recommended to safeguard, protect, and revive this system which includes two types of heritage: a tangible heritage which is the foggara (well, underground gallery and comb (kasria)) and intangible heritage which includes the knowledge and skills of men of the foggara.

REFERENCES

ANRH. (2012). Measurement of foggara flow in Adrar province. Adrar: National Agency of Hydraulics Resources.

ANRH. (2019). Foggaras Inventory. ADRAR: National Agency Of Hydraulics Resources. An internal report of the National Agency of Hydraulics Resources (in French Agence Nationale des Ressources Hydrauliques ANRH), Adrar, Algeria.

Ismaili A. A. (2018). Water Distribution Mechanisms in Touat. Their Origination and Development. Dar el Kalima, Adrar.

Mohammed B. (2016). 'A reading of aspects of the ancient architecture of Tamantait Palace through the manuscript of The Simple Speech in the News of Tamentit by Sheikh Muhammad al-Tayyib bin Haji Abd al-Rahim, known as Ibn Baba Haida'. *Rufuf*, **421** (5676), 1–15.

Nasri B. (2018). The water quality of Foggaras in Adrar province. ADRAR: National Agency Of Hydraulics Resources.

Observatoire du Sahara et du Sahel. (2003). Systeme aquifere du Sahara septentrional, 2ime Edition. Finzi, Tunis.

Professor Ahmed Djaafari (2009). 'For the historical and legal dimensions of The Book of the Questionnaire on what Happened to Tuat of Matters and Issues.'

Remini B., Achour B. and Kechad R. (2010). 'La foggara en Algérie: un patrimoine hydraulique mondial'. *Revue des Sciences de L'eau/Journal of Water Science*, **23**(2), 105–117.

World Health Organization. (2006). 'Water & Culture the International Decade for Water 2005–2015'. Available online at: https://www.who.int/water_sanitation_health/Water&cultureEnglishv2.pdf (accessed 5 January 2021).

Chapter 8

Customs related to water and water management in the tradition of the Dai people in Vietnam

Nguyen Thi Hien and Vu Hong Thuat

ABSTRACT

This chapter is an ethnographic account of the roles of water in the life of the Black Dai people in the Quy Chau and Que Phong Districts, Nghe An province in central Vietnam. For the Dai people, water meets the needs of their daily life, which includes housing, agricultural production, cultivation, and their customs and religious practices. This account relies on data that was collected for the case study in 24 villages of six communes in the two districts in 2013–2017. Using the views of local people and the information that they shared, this chapter demonstrates how the Black Dai people exploit and use water in an effective and efficient way. Their knowledge, their customs, and their practices reveal cultural identities that have been passed down by elderly people. Water is also reflected in their ways of social behavior and community management. Today in the open market economy and high technology era, their customs and habits related to the use and management of natural water endure and are still being promoted in their daily lives, such as in agriculture and swidden land cultivation.

Keywords: Black Dai people, water, customs, traditions, local knowledge, management

8.1 INTRODUCTION

The role of water in the lives of people living in areas with diverse natural conditions is crucial, leading to rich traditions related to the use and management of water. The

Editor: Xiao Yun Zheng
doi: 10.2166/9781789062045_0123

Dai people in Vietnam have their own traditions and customs related to water. Apart from meeting people's everyday life needs (eating, drinking, cleaning, for livestock and plants, etc.), water also has its own cultural and social meaning for the Dai people. This chapter demonstrates the different ways that the Dai use water for their daily life and for protection of their surrounding environment.

Through the case study in 24 villages of six communes in two districts of Que Phong and Quy Chau in Nghe An province, we show that the Black Dai minority people have maintained their customs and rituals related to water. In the life of the Black Dai people, tangible cultural elements, such as houses, clothing, transport, agricultural tools, and intangible elements, such as the practice of rituals, festivals and folk performance, and games, have changed in the context of globalization and industry 4.0. In contrast, customs, traditions, and rituals related to the management and use of water continue to be practiced, despite the modern technologies of using electronic pumps. The Black Dai people exploit and use water in an effective and efficient way: this has resulted from their local knowledge, enriched by the elderly people and shamans, and has been handed down from generation to generation.

Consequently, from the perspective of local knowledge and with the participation of the local people, this chapter provides an ethnographic description and a comprehensive and in-depth insight into the customs of using water through rituals in the family and community. It also highlights the way of managing and sharing water in agriculture. These practices are closely related to the spiritual life of the Black Dai people. Furthermore, certain points will be analyzed in the way Black Dai people observe, change, and maintain their customs in Nghe An province. Based on data obtained from the field trips to 24 villages of the Black Dai people, the chapter highlights the rich local knowledge of these people. This is related to the sustainable use and management of water, in contrast to the customs in the use and management of water by the Viet majority people in the lowlands, which are more like 'robbery' from nature and are unsustainable.

8.2 THE DAI PEOPLE IN NGHE AN PROVINCE

According to the results of Vietnam 2009 Population and Housing Census, the Dai people formed the third largest population (after the Viet and Tai people) among the total of 54 ethnic groups in Vietnam with 1,550,423 persons, making up 1.81% of the population nationwide. The Dai people call themselves *Táy, Thăy,* or *Tày*, and their official name is *Dai*. Before the 1990s, they mostly resided in the provinces of the northern part of Vietnam, such as Son La, Lai Chau, Dien Bien, Lao Cai, Yen Bai, Hoa Binh, and the central provinces of Thanh Hoa and Nghe An. The migration which started at the beginning of the 1990s extended their living areas to other provinces of the Central Highlands, such as Gia Lai, Dak Lak, Kon Tum, and Lam Dong (Nguyễn Công Thảo, 2016, p.122). The total population of Nghe An province in 2009 was 2,912,041, among which the Dai people made up

295,132, accounting for 69.92% of the total population. Thus, they live mostly in Con Cuong, Tuong Duong, Ky Son, Quy Chau, and Que Phong districts (Vi Văn An, 2017a, p.22).

The folk knowledge of water of the Dai people, in general, and the Black Dai people in Nghe An province, in particular, has been formed in close connection to their understanding and experience in dealing with the water environment. Therefore, this chapter addresses the knowledge that is related to customs in the exploitation and use of natural water sources in their daily life.

8.3 THE ROLES OF WATER OF THE BLACK DAI PEOPLE

8.3.1 Water as an important criterion in selecting a place to live

One of the most important criteria of the Dai people when selecting a piece of land to build a house is that the land should be near a water source. However, in recent years, in certain areas near towns, the Dai people have been paying attention to a new criterion – the proximity of roads, markets, and schools. Closely related to this is a new tendency of moving from their stilt houses to houses at ground level. In both cases, either living in towns or near water sources, most houses of the Dai people are situated together in basin valleys, and their villages are always close to a stream, river, water source, or waterfall. This makes it easier to install water pipes to supply water to their homes for cooking and cleaning, and for livestock, fishing, and for irrigating the rice fields (Figure 8.1).

Looking at the residence patterns of the Dai people, it is seen that many village names start with or have elements related to water (*nặm*), for example, *huổi* in the Dai language which means stream, *bó* (water source), *xốp* (stream mouth), *nong* (pond, lake), and *châu* (river/big stream). Examples of such village names in the northwest, as well as in other districts in Nghe An province, include Nặm Giải, Nặm Nhóng, Nặm Rốn, Nặm Xan, Nặm Xuông, and Nặm Bóc. Villages that are near rivers or large streams, which play a central role, usually have the name of *Tá Đo* (wharf). However, in practice, not all Dai people/villages live by cultivating rice in paddy-fields. Many of them live by combining cultivating rice in both paddy-fields and swidden land. Villages of the Dai people are always near water stream/water source which is convenient for bringing water home, using water for fishing, collecting reeds from rivers, finding snails, frogs, eels, and mice.

The Dai people believe that if there is a water source within a family's land, it is wealth granted to the family by heaven and the family will be rich. During the field trip to Quy Chau district in 2014, a visit was made to Mr. Vi Van Bao's house which was on a piece of land with a water source. According to him, in the customs of the Dai people, a house built on a piece of land with water source or near the upstream is believed to have good *feng shui*, the family would have good luck, children will be

Figure 8.1 The Dai people settle down at the foot of mountains by streams to gain water to irrigate for rice cultivation, Nam Nhoong Commune, Que Phong District, Nghe An Province, 2016. (With permission with VU, Hong Thuat).

well-behaved, husband and wife would be happy and would have both sons and daughters. More especially, they would be lucky in cultivation, husbandry, and fishering. Since Mr. Bao's house is on such a land, he is the wealthiest person in Xop Hoc hamlet. (Interviewed Vi Van Tien in August 25, 2014.)

By traditions and custom, the Dai people prefer to select pieces of land in the shape of a dragon or turtle and in front of the house there should be a stream or river. The belief is that if there is water, good luck will come. Also when a village is near to a mountain, it is believed that with *feng shui*, houses/families will have good luck. There are many similarities between the viewpoint of water sources and *feng shui* of the Dai people, as well as elements in the culture of the Viet people and many other ethnic minority groups in the north of Vietnam. In rituals, to pray for good luck and prevention of bad luck, water and fish are needed. The water culture of the Dai people is very diverse, and is unlike the culture of the Việt people in the lowlands.

Observing the house of shaman Lu Trong Bang, it was seen that he had two fish ponds in front of his house, each about 500 square meters. He is a shaman who sees in depth that the surroundings of the Dai people would be associated with water. Mr. Bang's statement gave a better understanding about the two ponds. He said that: 'We believe that water source is the same as gold source. Because my family has decided to build our house on a piece of land in the form of a turtle puffing out water, I dug those two ponds as a symbol of two water sources with the meaning of harmony between *yin* and *yang,* land and heaven, humans, and gods.' (Interviews with shaman Lu Trong Bang, Quy Chau district, Nghe An province, August 2014.) Shaman Bang's choice of a piece of land associated with water to build his house on and the fact that a village usually settles near a water source demonstrate an important role of water not only in agriculture and everyday life but also in their spiritual life. His house is sited according to the *feng shui*. His expectation is to be the wealthiest in the commune and that his family members will have good livestock, and be healthy and happy.

8.3.2 Water serving the needs in production, for livestock, and source of food

As an agricultural population that majors on cultivating wet paddy rice, the Dai people have gained a lot of experience and local knowledge in the exploitation, preservation, and effective use of water resources to serve their needs in production and husbandry. Their experience and local knowledge have created part of their identity values in the west of Nghe An province. Traditionally, the Dai people in Vietnam are usually known as inhabitants that 'live with water'. This is an image, implying that the Dai people have developed cultivation of paddy rice. Thus, this plays an important role in producing the main source of food, which is different from many ethnic groups living in mountainous areas and relying on the cultivation of swidden land. This viewpoint has some similarity with that of the Viet people that in agriculture, water plays the most important role (first – water, second – fertilizers, third – hard work, and fourth – breed) for high productivity. The big rice fields at the foot of mountains or along rivers and streams are results of exploitation by many Dai generations through the centuries. The Dai people are said to have the most diverse livelihood forms among the community of ethnic minority people in mountainous areas in Vietnam (Figure 8.2).

Descriptions of the traditional water rice agriculture of the Dai people in Vietnam emphasizes the traditional irrigation system of *mương, phai, lái, lín* and *cọn nước* (water-wheel). In addition, the Dai people are regarded as clever in taking advantage of natural characteristics to provide water for crops. With the irrigation system of *mương*, *phai*, *lái*, and *lín*, a small *phai* (dam) is built across a stream/river to raise the water level, thus guiding the water flow to a *mương* (ditch). Also, from the ditch, water is divided into smaller flows (called *lái, lín*) and it flows into rice fields. All of this is called a "water guiding into fields system" (*dẫn thủy nhập điền*) and allocates water to families in the villages. Each family has a separate field to cultivate rice and they are able to two crops every year. In the first or second lunar month, the first rice crop is planted and harvested in the fourth and fifth lunar month, while the second rice crop is

Figure 8.2 The irrigation distribution of water from the upper terrace to the lower terrace before the rice cultivating season, Nam Nhoong Commune, Que Phong District, Nghe An Province, 2016. (With permission with VU, Hong Thuat).

planted in the fifth or sixth lunar month and harvested in the ninth or tenth lunar month (Vi Văn An, 2011).

Although in the past they used only manual labor, the Dai people achieved substantial results in construction dams and ditched for their irrigation system. There are still many ditches in use today in the Northwest and Nghe An provinces more than 1 km long, going along hill slopes and passing barriers to bring water to the fields.

The *phai* is a dam made of wood, bamboo, straw, soil, or stone to increase the water level, thus guiding water to a ditch, from there to fields or water wheels and to further pour it into bamboo pipes and then into fields. The importance of *phai* is expressed in the saying 'when a *phai* is broken, the worry is the same as when father dies' (*pó tai khư phai lớ*).. This, however, depends on the actual conditions where a particular *phai* is built. The ditches and *phai* are the most important elements to ensure rice productivity, so customs of the Dai people stipulate that every member of the village/community is obliged to participate in building and protecting the ditch and *phai*. Also, they have the right to use water brought into their fields. In each village, this work is usually managed and directed by a chief. The chief is also responsible for appointing villagers to repair the *phai* as well as guard the ditch and *phai* during the rainy season.

Lái is a type of additional *phai*, which is usually built across a ditch to guide water to small rice fields or guide water flow into a water-wheel.

Lin is the system of pipes guiding water from the water-wheel to the rice fields, and it is usually made of different types of bamboo or palm trees or other types of tree in the form of a hollowed-out trunk (nowadays in some areas, PVC pipes are used instead of traditional materials). In other words, *lin* functions similarly to a ditch in guiding water to the fields. If there are barriers or if the field is too far away, it is very difficult and impossible to dig a ditch, the *lin* is a better way to bring water to rice fields. Therefore, the *lin* is usually put on a system of strong supporting frame, made of bamboo. Two bamboo poles are positioned cross-wise, and then the *lin* is placed on the angle.

In many places, there is a distinction between *lin* and *máng*. If the pipes are made of hollow tree trunks or palm bark, the system is called *máng*; if it is made of different types of bamboo, it is called *lin*. However, both *lin* and *máng* have the function of bringing water to a ditch and then to rice fields. Depending on the water flow, sometimes it is necessary to create *máng* from one bank of the stream to the other to guide the water flow to the ditch, then to rice fields. In this case, every 5 to 6 meters, a cylindrical bamboo basket (*sọt*)is made of diameter about 1.5 to 2 meters. The *sọt* is placed upright, while two poles are placed inside and stones are packed into the basket to hold the poles upright. The two poles are secured with bars and serve as a frame to support the *máng*. Therefore, this way of building *máng* across a stream is also the initiative and experience of the Dai people in irrigation.

In addition, the water-wheel (*cọn nước*) is another unique cultural invention in the irrigation system of ethnic minority groups such as the Tày, Mường, and Dai. It has demonstrated not only their behavior in an agricultural environment but also their achievement and experience in the field of traditional irrigation – the calculation of water flows to find suitable places for water-wheels (Vi Văn An, 2011). Water-wheels are constructed from bamboo and wood. The wheel uses a fast-flowing stream to create a force to make the wheel rotate and lift water and pour it on to higher-level fields. As the wheel rotates, bamboo tubes automatically take water and pour it into a ditch or *máng* so that water continues to flow in the ditch to the rice fields. Therefore, for the wheels to rotate, it is necessary to build a small *phai* and dig a ditch or V-shaped opening. This creates a fast flow that is sufficient to force the wheels to rotate. Bamboo tubes take water from the stream, bringing it to the top of the water-wheel. From there water goes to the *máng* or ditch.

In general, a water-wheel is made from two main components: the wheel axis called *pặt* and two supporting poles called *mé pặt*. The rotating axis is made from a wooden core (*lay*) of a type of strong wood, such as *vàng dành, vàng tâm, táu, trai,* or *sến*. Bamboo pipes of equal length are tied to the two sides of the axis (*xào*) to create a series of triangles, which are attached to a ring made from bamboo or creeping plants called *khớp*. Two inner bamboo rings are used to strengthen the wheel and ensure that when it operates, the bamboo pipes do not move apart. Bamboo tubes called *boing nặm* are tied to the circumference of the wheel. As the water-wheel is being built, a supporting frame is also constructed.

The supporting frame is made of two poles set at the same distance apart as the length of the rotating axis. These two poles might have a natural and hand-made *ngoãm* to create a square where the axis will be placed. If it is not convenient to bury these poles in the ground, they are put into bamboo baskets and kept firm by stones packed into the baskets. Depending on whether the fields are high or not, or the stream bed is hollow or deep, the poles are made tall or short. The diameter of the water-wheel also depends on these factors (Figure 8.3).

Building a water-wheel is a difficult and time-consuming task, requiring experience in location selection. Also, it involves the calculation of water flow at different and critical times (for example, when there is heavy rain or flooding) to see whether it is necessary to build an additional *phai*. However, a water-wheel is not built anew for every crop/season, thus materials for a water-wheel are usually carefully selected so that they will last for a long time.

In helping each other during the construction process as well as dealing with issues which might arise later, such as when floods destroy a water-wheel, villagers usually build water-wheels on a certain day, usually on an even day (2, 4, 6, 8, etc.) on a piece of land beside a river. Therefore, water-wheels of a village usually stand next to each other. This creates a cluster of water-wheels which is pleasing to the eyes and produces nice sound when they are in operation. This cluster of water-wheels is known by the Dai people as *pại pặt pừng*.

The local knowledge of the Dai people is applied very effectively in their use of natural water sources. They employ wate-wheels to bring water high to pour into

Figure 8.3 The Dai people make water-wheels for irrigation, Dien Lam District Que Phong District, Nghe An Province, 2016. (With permission with VU, Hong Thuat).

ditch, and *máng* has proved that the Dai people have solved the issue of saving manpower for building *phai*. This involves digging hundred meters of ditch through many barriers which is very difficult by manual labor in mountainous areas. The water-wheel which is so successful in raising water to supply the rice fields is a material structure characterising agriculture in the valleys at the foot of the mountains.

The Dai people in Vietnam in general and in the west of Nghe An province usually exploit two sources of water for agriculture – water from big streams (for fields with two crops per year) and rainwater (for fields with one crop). Every year, before plowing, each family appoints one member to participate in cleaning ditches and the *phai*. When there is rain and flood, nearby families are responsible for closing the ditch mouth to prevent too much water pouring into rice fields. If the *phai* is destroyed, then the whole village will have to repair it. After a flood, rice fields usually become dry; thus cleaning the ditch at that moment is also a task for the whole village. About ten days before harvesting the crop, the Dai people usually empty water from the fields, making more convenient for gathering rice and, at the same time, catching fish in the fields and collecting straw. The task of emptying water is also clearly communicated to every family.

The Dai people in Vietnam cultivate rice in terraced rice fields. As such, the highest terrace of rice gets water from the ditch or *phai*. After then bamboo tubes are put through the small bank, connecting water from higher terrace to lower terrace. In this way, the farmers save water and divide water between fields or families, avoiding disputes over water in the community. In this way, the Dai people have achieved significant results in exploiting and using irrigation water. The Dai people have always considered irrigation to be very important role, as shown by an expression in the Dai language: 'there should be water to make fields, there should be fields to have rice' (*mi nặm chẳng pên na, mi na chẳng pên kháu*) (Vi Văn An, 2011).

The division of water from a ditch to fields in each village is managed by the village chief. For example, in *Tằm I* and *Tằm II* villages, Quy Chau District, families having neighboring fields are not allowed to dig an opening on their own to bring water to their field. They are expected to follow the principle of letting water flow into higher fields first, then into lower fields. On the contrary, for families with separate fields, building *phai*, digging ditch, and guiding water to their fields are all done on the families' own initiatives.

It can be seen that water is not only a criterion for choosing a place to live or to build a village, but it also plays an important role in exploiting fields and creating conditions for the formation of dense population groups. Areas where the Dai people have lived for a time long in the Northwest and the West of Nghe An all forms the basis of big fields. An expression in the Dai language states: 'There should be water to be ditch, there should be a village – *mường* – to be a chief' (*mi nặm chẳng mi mương, mí mường chẳng mí tạo*). Thus, this means water is the first condition. When there is water, people can then dig a ditch, a village should be built first, and then a chief selected to manage the village. The implication of this expression is the role and importance of water as an inherent factor in the social formation of their villages (Vi Văn An, 2011).

For the Dai groups living mainly by cultivating on swidden land, rainwater plays an important role in cultivating rice. Rain is the main source of water for cultivation, thus the growing season for rice and other plants (vegetable, beans, groundnuts, chilli, etc.) is associated with the rainy season. According to the agricultural calendar of the Dai people, clearing and preparing the swidden fields of the Dai people takes place in the third and fourth lunar month, burning and sowing seeds in the fifth lunar month, destroying grass in the sixth and seventh lunar months, and harvesting in the ninth and tenth lunar months.

Using local knowledge, the Dai people always understand the rules of the weather (for example, sunny and rainy days) and they take advantage of the natural source of water in agriculture in the most effective way. For example, the Dai people usually end seed sowing by the end of the fifth lunar month so that the rice can grow to the height of about 20 centimeters to catch the first rain showers at the beginning of summer. Rainwater infiltrates through the surface of the ground and reaches the first layer of the ground with fertilizers, which helps the rice to grow faster. Mr. Tien comments: 'During the process of agricultural production, we should gain experience of recognizing weather to know how weather is changed from year to year. As the folk experience, if the New Year's Eve is very dark, that year will have much rain, people would cultivate on swidden land, if the New Year's Eve is lighter, there won't be much rain that year and it would be better to plant corns, beans and cassava, etc.' (Interview with Mr. Lu Van Tien, Quy Chau district, Nghe An province, August 2014.)

Local knowledge, based on the effective use of rainwater to ensure rice growth, can be seen not only in rotational cultivation (meaning growing one type of plant, then changing to another type, and coming back to the first), but it can also be seen in the way land is selected to cultivate rice on. Mr. Bao says that 'The most popular way of selecting land to be burnt and cleared for rice fields is to firmly put a knife into the ground and take it out; if soil sticks to the knife, it means the soil has enough moisture. If no soil sticks to the knife, that land is too dry, and can't be the land to cultivate on.'

The irrigation system of the Dai people in Vietnam, in particular, and in the West of Nghe An province, in general, is a valuable example of folk experience and knowledge. Their method of irrigation demonstrates their adaptability to the natural environment and their knowledge and experience has been accumulated through many generations of agricultural inhabitants. The irrigation system of the Dai people demonstrates not only a common identity of the water rice civilization of Southeast Asia, but it also has a specific identity of the Dai people in Vietnam and is a sound technical solution in the exploitation of natural sources of water of this ethnic minority group (Lê Sĩ Giáo, 1992, pp.23–30). The irrigation system is not only evidence of the high level of the local community's knowledge but is also symbolic of ethnic culture and social relationships.

The Dai people use plots of land around their houses to dig ponds for fish culture and husbandry to have food for their New Year holidays, when guests come to their houses, for rituals, and for selling produce to other people in their villages. More than 50% of households in the two study districts have fish ponds. These fish ponds are usually small, with an area of 20 square meters to 100 square meters. The fingerlings (very young fish) are provided by the Việt people, brought from the lowlands, and sold in the third lunar month. The fish are then cultured until the end of the year to be used for the New Year holiday. In the cuisine of the Dai people, grilled fish has become a cultural feature and during the Tet holiday, grilled fish must be presented as an offering to the ancestors.

In addition, the natural sources of water in streams, rivers, and ponds has provided the Dai people with abundant food. During their leisure time or after each flood, women go along the stream or river bank and use a trap to catch fish or shrimp and fetch weed to eat as a type of vegetable; men go fishing or cast their nets to catch fish. When there is heavy rain, water floods fish ponds and the whole village goes out fishing. However, when the water level is low, people use nets to catch fish or find crabs and snails. In most houses of the Dai people living along streams and rivers in Nghe An, there are fish nets and fish traps. Freshwater is also a habitats for high-value belostomatids (giant water bugs) and amphibians, such as frogs. In the diet of the Dai people, seafood is an important component. Fish, crabs, and snail are the second or third important element (after rice and vegetable), just like the Việt people and other ethnic groups. Fish can be prepared in different ways: grilling, frying, steaming, boiling, as porridge, etc. One of the favorite dishes of the Dai people is *nậm pịa* (a kind of fish sauce) (Trần Ngọc Thêm, 1998, p.66).

Water is also a means of transport for the Dai and many other ethnic groups living along big rivers such as the Black River, Lam River, Chu River, and Ma River. They use canoes or rafts to transport people and goods across rivers and streams or for fishing. Hence, this has become the second important cultural symbol after stilt houses (Vũ Hồng Thuật, 2016) (Figure 8.4).

8.3.3 Water for everyday life

Besides its use in watering the fields and for livestock and plants, water is also a natural resource necessary in the everyday life of the Dai people. They use water for cooking, drinking, bathing, cleaning and operating rice mortars, and producing electricity by hydropower. Most of the Dai households in communes in Que Phong and Quy Chau districts currently use water for domestic purposes from streams, rivers, or water source in the forests. Depending on the area and source of water available in nature, the customs in water exploitation and use differ from place to place (Figure 8.5).

Figure 8.4 The Tai people make use of stream flows to carry wood from forest to home, Dien Lam District Que Phong District, Nghe An Province, 2016. (With permission with VU, Hong Thuat).

Water for bathing, cleaning, and eating is taken from streams. This way of using water is very common in all villages of the Dai people in Vietnam. Bathing, cleaning, and washing clothes usually take place outside – in rivers, streams, or at the end of the water source or by the family's water tanks. In the past, if water was taken from water source, it was taken in the early morning, at noon, and in the late afternoon. Water was brought home in bamboo pipes and then poured into wooden containers. Nowadays, water is brought home from water sourcse and streams in forests using PVC pipes to cement tanks, helping people to save time. In many places, people distinguish between water sources for bathing and laundry, and water source for eating and drinking. Water for eating and drinking is mostly taken from underground water or upstream in forests which are more hygienic than water taken directly from streams or rivers. To have a source of clean water, in villages of the Dai in Chau Phong commune and Quy Chau district, the Dai people clean and build an embankment around a headwater pond and construct a system of pipes to bring water home. Water for everyday life can be taken from a source used for the whole village. Also, several households take water from one source or each household takes from their own source.

Nowadays, due to the communication, media, and community healthcare campaigns about water use – 'eating cooked and drinking boiled' – the Dai

Figure 8.5 The Dai people bathe and wash clothes in streams, Dien Lam District Que Phong District, Nghe An Province, 2015. (With permission with VU, Hong Thuat).

people in many places have changed their habit of drinking natural water to drinking boiled water. In the two districts in Nghe An province, the field surveys show that the Dai people in hamlets have received funds from the Government and organizations such as UNICEF, PAM, and Oxfam to build water tanks for the whole village. However, after several years, the system does not work anymore due to overexploitation of the water source (use of one big pipe for the whole village and this has led to the shortage of water to be brought to the water tank). In contrast, the old local way of using water in which two to five households share a water source proves to be more effective when water slowly flows into the pipes to be brought to households (Figure 8.6).

Water from streams was also used for operating rice mortars (before the national electrical grid was constructed and there was no machine for rice milling). A shelter covered with palm leaves was built for the rice mortar next to the river/stream. In addition, a *phai* (small dam) was built to guide the water flow to take advantage of the water pouring into the trough at the tail of the mortar, causing the mortar to operate all the time, thus cleaning the rice husks. Early in morning when they went to the fields, they poured whole rice into the mortar. In the later afternoon, returning from the fields, they took rice which had been removed from its husk. This helped to save human labor. Nowadays, the villages of the Dai people are connected to the national electrical grid and

Figure 8.6 The Tai people use cans to take drinking water from forests to home, Nam Nhoong Commune, Que Phong District, Nghe An Province, 2016. (With permission with VU, Hong Thuat).

many households have their own rice mini-milling machines and do not use the traditional rice mortar any more.

For the last 20 years, the Dai people in the west of Nghe An province have been using small electricity generators (made in China, capacity of 0.5–1.5 kilovolt). People take advantage of the water flow from streams and rivers, build small dams, and guide the flow to the generator to produce electricity which is used for lighting, radio, video, TV, fans, etc. Though at the time of this study (2016), villages of the Dai people in the two districts had been connected to the national electrical grid, small hydropower still plays an important role in peoples' lives. Many families use both sources of electricity to reduce the costs (Figure 8.7).

8.3.4 Water in rituals and magic

The Dai people in Vietnam have a very diverse and rich culture. In this wealth of culture, water plays an important role. It is not only a rich resource but also a foundation and basis for the Dai people to create their unique cultural identity. Being a population living and farming in association with water, the Dai people in the west of Nghe An have many customs and habits which are closely related to water. Water has become sacred and has not only has the power to send away ghost and devils, it also has the power to protect humans, plants and livestock,

Figure 8.7 In 2000, UNICEF supported financially to buil water tanks to contain water coming from streams in forests, Nam Nhoong Commune, Que Phong District, Nghe An Province, 2014. (With permission with VU, Hong Thuat).

clean, destroy and revive humans when they die and go to the other world. Water has also become a symbol and belief of the Dai people in their spiritual life.

Through field trips and interviews with the Dai people, much material related to their customs in the use of water were collected. During the Tet holiday, the Dai people clean the house, kill a pig, make cakes, and plan all the chores for welcoming the New Year. They also bring houseware items to the wharf to clean, which implies washing away bad luck in the old year to receive good luck in the New Year (Vu Hong Thuat, 2015a). They usually observe this custom at sunrise, in accordrance with the belief that the period of 7–9 am is the hour of dragon and bad luck would be taken away by the flow of water. Shaman Lu Trong Bang explains: 'The Dai people believe that dragon is sacred. It lives both in the sky and in water (river, stream, and pond), thus it can bring both good luck and bad luck. At sunrise, the dragon is living in water, thus it will carry away all bad luck.' (Interview with Mr. Lu Trong Bang, Quy Chau district, Nghe An province, August, 2015.)

Another Tet holiday custom of the Dai people, that takes place at the village wharf, is the ritual of calling spirits (*hiếc khoăn*) for family members. In a Dai family, the wife plays the most important role. She takes care of everything and is responsible for the kitchen, plowing, planting, harvesting, and storing of seeds. On the morning of the first day of the New Year, wives get up early to collect water from the head of the water source so that it is clean and fresh – this is called water of the New Year (*nặm pi mớ*) and ensures the family gets good luck for the year. This is an important custom, demonstrating the way people treat natural environment and their beliefs about water in everyday life. When she comes home, she takes three bamboo cups of the water to place on the altars for the ancestors and for the tutelary god, believing that by doing this, the ancestors and tutelary gods will protect the family and bring good luck and health to the family members, plants, and their livestock. After that, she gives each family member a sip of water to drink as a wish for good luck and health in the New Year, sending away bad luck and sickness. At the same time, family members use the water to wash their faces and clean their hands and feet. If a family does not go for water on the first day of the year, it is believed that the family will not have good luck for that year. This custom emphasizes the role of water in people's life and educates young people to protect and guard the source of water – for worshipping their ancestors and for use in agriculture.

In rituals related to the life cycle, there are also customs related to water. During pregnancy, Dai women avoid bathing in rivers and streams, fearing that the dragon god, snake god, or water god might bring bad luck to their unborn babies. After giving birth, the clothes of the mothers and their children are not taken to the rivers or stream for washing to avoid polluting the water. If the water were to be polluted that way, the water god might be angry and punish them by causing the child to be sick. More especially, the Dai people have a taboo of not urinating, defecating, and spitting into the rivers and streams, fearing that the water god might be angry and cause flood or draughts.

When a girl grows up and gets married, during the ritual of receiving the bride, the Dai people have the customs of accompanying the bride home when it is still dark to avoid ghosts and devils seeing her beautiful face. With such a belief, when a bride comes to the bridegroom's house, she stands for some moments by the foot of the stairs and the mother-in-law pours a small bucket of water which has been blessed by a shaman on her feet to send away ghosts and devils. This custom has a spiritual meaning that the water from the head of a water source is sacred and has the power of sending away ghosts and bad luck. It is also an expression of the social relationship that since from that moment on, the girl has officially become a member of her husband's family (Vũ Hồng Thuật, 2017).

As a life rule, from the moment of birth till death, human beings need water. When someone is sick, the Dai people believe that this is caused by bad things done by their ancestors in their previous lives and now their children are being punished by the gods (Vi Văn An, 2017b, p.148). From the spiritual perspective, a common way of explanation of sickness of the Dai people is based on animism, namely, human beings survive because of their souls (*khoăn*/*văn*) and *vía* (*ngau*) living in the body. As shaman Lu Trong Bang explains, 'One of the characteristics of soul is that it likes wandering, and might forget the way home or be kidnapped by ghosts and devils, not be able to come back to the body, thus a person becomes sick. When a person is sick, it is necessary to perform rituals to get the soul back and these rituals are associated to the customs of praying for good health.' (Interview with Mr. Lu Trong Bang, Quy Chau district, Nghe An province, August 2015.) Therefore, in the rituals to return the soul, in addition to giving water as offerings to the gods and the soul of that person, shamans also ask the person who has lost his soul to go to the river to bathe and wash the costume worn during the ritual. Therefore, this implies sending away all bad luck and praying for good luck from the gods.

In the death rituals, when a person just stops breathing, his/her relatives take water to rub the deceased's face and wish that his/her soul will go to their ancestors. After this they begin to prepare water with herbs to bathe the dead before putting the body into the coffin. Beside rituals and customs of water associated with both the living and the dead, the Dai people also use water in magical rituals to send away ghosts and devils and cure disease. Rainwater or water taken from rivers and streams is used in these various rituals. A shaman reads spells to water bowls, turning an ordinary water bowl into a bowl of sacred water, bearing the color of magic and religion (Vũ Hồng Thuật, 2015b, p.406–413).

To ensure they have water for production and everyday life, as well as prohibitions on destroying the forest upstream, the Dai people have a ritual associated with agriculture – the ritual of praying for rain, especially during a drought. Usually, when the community organizes worship rituals at the beginning of the New Year, shamans pray for favorable weather, a good harvest, and good health for people and livestock, etc. Often, the prayer for rain is combined in such rituals to reduce costs of offerings and to save time too.

In the customs of the Black Dai in Pog village (Que Phong district), there are clear rules related to offerings for the ritual worshipping of the hamlet god: 'There should be a rooster, a duck, a pig, sticky rice, and wine. The pig is for the forest god, the rooster is for the hamlet god, and the duck for the water god. These offerings are taken from the community fund donated by the community members, and after the ritual all the people gets their share.' (Private talk with the shaman Vi Va Thuong, Que Phong district, Nghe An province, August 2015.)

This is also an element demonstrating the relationship between humans and gods and the natural environment regarding their exploitation of natural resources in the forest as well as in rivers/streams. Here, the Dai people pay special attention to water resources and pray to water gods. Based on their belief, *phai* is the habitat of the river or stream 'owner' (*chầu nặm*). The river owner is symbolized in the image of a *thuồng luồng* (*tô ngược),* a creature in the same family as dragons. According to shaman Lữ Trọng Bằng (Quy Chau district), 'the stream 'owner' – *thuồng luồng* is the god to decide the existence of the *phai,* thus for the *phai* to be strong, unbroken by flood, every year people organize rituals to worship *phai* god, water god – *thuồng luồng* – in the eighth lunar month or combine this ritual with the worship ritual for the village god at the beginning of the year.'

The ritual to pray for rain was organized beside a wharf, led by a shaman-widow (*me mải*) together with a group of children. A hole was dug and a piece of areca tree used to cover the hole. An areca leaf (symbolizing a dragon) was inserted into the hole, and then the shaman-widow and children pulled the leaf toward themselves. The children run to the stream and splash water over each other, implying teasing the dragon so that it will not be angry but will cause the rain to fall (Figure 8.8).

Figure 8.8 A shaman performs rituals for abundant water and harvest, rain and wind, Nam Nhoong Commune, Que Phong District, Nghe An Province, 2016. (With permission with VU, Hong Thuat).

8.4 CUSTOMS RELATED TO WATER MANAGEMENT

Recognizing the role and importance of water in human lives, since ancient times, the Dai people had clear rules to protect this valuable natural resource. These rules are clearly expressed in their customs and habits (*hịt khoong*) related to the two aspects – water in everyday life and water for agricultural production. This is the way of thinking and it reflects the relationship between human beings and their environment.

In the customs of the Dai people, whether they cultivate rice in the paddy-fields or on swidden fields, water is still important for the livestock. The Dai people understand that if there is shortage of water, crops will die or produce will be poor. Therefore, the traditional society has clear rules related to the protection of water sources that are well expressed in the customs or rituals to worship the village god, river god, water god, and agriculture god every year. For the Dai people, their society is divided into cantons *(mường)* and each of the *mường* consists of many villages and hamlets. In traditional society, the Dai people were divided into nobility and civilians. In theory, land, water resources, forests, and natural resources belong to the whole community (*mường*) and the land owner (*chẩu mường*) is the community representative who takes responsibility for management.

Beside folk knowledge of using and managing natural resources in general and water in particular, in the customs of the Dai people, forests are usually sacred by being associated with magic stories related to gods bringing to people good luck or bad luck and also associated with taboos. Those who dare to violate those taboos, by making the gods angry, will be punished. For example, according to the Black Dai people in Quy Chau district: 'In each village, there are gods (land god, forest god, river god, village god, etc.) who bless families in the village. Gods reside in pieces of river/stream with much fish, or waterfalls, water sources, forbidden forests. Thus people are prohibited to urinate or defecate into the head of the rivers, streams or in forbidden forests. At the beginning of the year, after the rituals worshipping the village god, mountain god, land god, and river god by the whole community, families can take water per household.' (Interview with Vi Van Tien, Quy Chau district, Nghe An province, August 2015.)

To protect the source of water for agriculture, everyday use and rituals, in each village, there are rules on the responsibility and punishment related to the violation of those rules. The rules (*hít khoong*) in *mường* Muỗi (Thuan Chau district, Son La province) stipulate: 'In *Mường* Muỗi, there is Khâu Tú mountain which is the mountain for worshipping ancestor *Ăm Poi*, the ancestor of the head of the village. At one end of the *mường*, there is a 'forbidden forest' called *cửa Xen* which is a sacred forest with many big trees. Next to the *mường*, there is also a forest called *Chiêng Kẻo* which is a ghost forest and it is also forbidden to cut trees there ….'

As for primeval forests, there are clear rules to prohibit cutting trees: 'There is land with forest where gods allow human to live. Regarding the forbidden forest at the upstream where ghosts reside, it is prohibited to destroy or cut trees down' (Ngô Đức Thịnh & Cầm Trọng, 2003, p.584). In many places, taboos on forbidden forests still exist in the community of the Dai people such as in the Northwest, the West of Thanh Hóa, and in the Nghe An provinces. These forests contribute to the preservation and protection of the environment as well as preserving the natural sources of water for agriculture and everyday life of human beings and livestock.

If water for production can be exploited and taken from different sources, water for everyday life should be taken from cleaner sources. Therefore, the community also pays more attention to the protection of the source of water for everyday life. Rules regarding the management of sources of water for everyday life use, especially water for eating and drinking, are also more specific and clear in the customs of the Dai people in each locality.

8.5 CONCLUSION

Knowledge of the customs of use and management of natural water sources of the Dai people in Vietnam, in general, and in Nghe An province, in particular, is valuable for studies of this ethnic group and their cultural identity and heritage, and it also reflects their social behavior and community management. In the market-oriented economy, different aspects of socio-economic life of the Dai people have been facing changes. In addition, new social concepts and forms are emerging. However, customs and habits related to the use and management of natural sources of water of the Dai people in many localities have remained unchanged and have kept their positive meaning. The main issue here is how to combine the two – local knowledge and scientific knowledge – in protecting, managing, exploiting, and effectively using natural resources, in general, and water resource, in particular, in a sustainable way and in the context of complex global climate change as at present.

REFERENCES

An Vi Văn (2011). 'Tri thức dân gian của người Thái trong việc sử dụng và bảo vệ tài nguyên nước' [Local knowledge of the Dai people in the use and guarding the water resource]. Tạp chí *Văn hóa Nghệ An [Journal of Nghệ An Culture]*, Sunday, April 17, **199**, 17–25.

An Vi Văn (2017a). Người Thái ở miền Tây Nghệ An [The Dai people in the West of Nghe An]. Thế giới Publishing House, Hanoi.

An Vi Văn (2017b). "Yếu tố ma thuật trong cách chữa bệnh của người Thái ở Nghệ An (dân gian và hiện đại) [The magic element in the treatment of disease by the Dai people in Nghe An (folk and modern]). Tạp chí *Nghiên cứu tôn giáo [Journal of Religious Studies]*, **5&6**, 146–161.

Giáo Lê Sĩ (1992). 'Các đặc điểm của nông nghiệp truyền thống của người Thái Việt Nam [Features of the traditional agriculture of the Dai people in Viet Nam]. Tạp chí *Dân tộc học [Journal of Ethnography]*, **1**, 23–30.

Thảo Nguyễn Công (2016). 'Dân tộc Thái' [The Dai ethnic group], in *Các dân tộc ở Việt Nam [Ethnic groups in Viet Nam]*, volume **2**, Tày-TháiKađai language group, Chính trị Quốc gia publishing house, Hanoi.

Thêm Trần Ngọc (1998). 'Vai trò của nước trong truyền thống văn hóa Việt Nam và Đông Nam Á' [The role of water in Viet Nam culture and in the culture of Southeast Asia]. Tạp chí *Văn hóa Nghệ thuật [Journal of Culture and Arts]*, **8**, 60–69.

Thịnh NgôĐức and Trọng Cầm (2003). Luật tục Thái ở Việt Nam [Customs of the Dai people in Viet Nam], Văn hóa Dân tộc Publishing House, Hanoi.

Thuật Vũ Hồng (2015a). 'Tết truyền thống của người Thái ở miền tây Nghe An' [The traditional new year of the Dai people in the West of Nghe An]. Tạp chí *Cộng sản [Journal of Communism]*, **1&2**, 17–21.

Thuật Vũ Hồng (2015b). 'The story of being a shaman of the Dai people in the West of Nghe An in the community of the Dai- Kadai people in Viet Nam and sustainable development issues'. Seventh national conference on the Dai studies in Viet Nam, 2015, Lai Châu, Thế giới Publishing house.

Thuật Vũ Hồng (2016). Thuyền độc mộc trong đời sống văn hóa các dân tộc ở Tây Nguyên và các tỉnh miền núi phía Bắc [Canoes in the cultural life of ethnic minorities in the Highlands and Northern provinces], project at Ministry level (2015–2016).

Thuật Vũ Hồng (2017). Tác động của ti vi đến mối quan hệ gia đình và cộng đồng người Thái ở bản Pục, xã Nậm Giải huyện Quế Phong, tỉnh Nghệ An, [The impacts of the television on the relationships between families and communities of the Dai in Puc village, Nam Giai commune, Que Phong district, Nghe An province]. Fieldwork report in August, 2017. Viet Nam Ethnology Museum.

Chapter 9

The role of water culture in water management: A case study on the mid-reaches of Red River, China

Xiao Yun Zheng

ABSTRACT

Based on the structure of water culture, which include the four levels of ideas, institutions, social life and physical life, this chapter surveys the dynamic role of water culture in local water management using a case study on the mid-reaches of the Red River in China. This study shows that local water management closely depends on water culture. The local people, based on their ideas, values and knowledge about water use in the local water environment, created an environmentally friendly irrigation and life system. This was also related to their religion, social norms, rules, and social relations and has made water management to be of significant importance. This case study also demonstrated that water culture is indispensable for water management to support the irrigation system. It is, however, closely related to local subsistence since the terraced wet paddy-field is the essential resource of traditional subsistence. Water management and subsistence will not be sustained without water culture. This chapter also discussed the future prospects of water culture in a fast changing society and focuses on how to maintain traditional water culture and build a new culture to meet the current demands of water management.

Keywords: China, midstream valley, Red River, subsistence, water culture, water management

Editor: Xiao Yun Zheng
doi: 10.2166/9781789062045_0145

9.1 INTRODUCTION

Currently, cultural factors are increasingly important for water management, water environment treatments, and responses to the challenges of global water environment changes (Xiao Yun, 2013a). Water culture has played an important role in water management in the past, and it is still valuable as an inheritance and is continually renewed. Mass cultural elements related to water have emerged in human societies from the distant past to present day and form an aspect of human culture known as 'water culture'. Water culture includes water-related ideas, myths, religions and social norms, customs, rules, laws, knowledge, and techniques of water use, management and treatment, artistic expression of people's understanding and impressions of water. Water is an indispensable substance for life and water management is necessary to sustain life. Therefore, from earliest times people worldwide had to find ways to manage water, starting from an interaction process with water. Water culture was generated as people interacted with water and has a significant influence on water management. Fundamentally, water management is a cultural process since it is based on the cultural factors mentioned above. Culture of water management is often neglected since it is difficult to define.

This chapter uses a case of the mid-reaches of the Red River, China, to survey water culture's role in water management from daily life to subsistence and in water sustainability. The data used in this chapter was collected in the fieldwork in 1999, 2009, 2011, and 2017 in the Yuan Yang and Red River Counties, Red River Prefecture of Yunnan Province, China. These two counties are the main portion of the Red River Valley in China.

The Red River rises in Weishan County, Yunnan Province of China. The Red River Basin covers Dali Bai Nationality Autonomous Prefecture, Chuxiong Yi Nationality Autonomous Prefecture, Simao City, Yuxi City, Kunming City, Honghe Hani and Yi Nationality Autonomous Prefecture, and Wenshan Zhuang Nationality Autonomous Prefecture. It flows through seven prefectures and cities in Yunnan and the area of Yulin County of Guangxi Zhuang Autonomous Region, and it flows into Vietnam across Lào Cai, Yên Bái, Việt Tri, Hà Nội, Hưng Yên, and Thái Bình. Subsequently, it also flows into Beibu Gulf (Gulf of Tonkin) via the Red River Delta. It is referred to as the Yuanjiang River in Chinese territory and as the Red River after flowing into Vietnam. In total, it reaches 962 km in length of which the Chinese section is 692 km, and the Vietnamese section 270 km (The Editing Committee, 1998). The basin was formed in different geographic landscapes as the river flows through different areas, and there are many nationalities living along the basin. Ecological diversity developed among people and the local environment, and many types of ecological culture were formed as people fitted into their local environment. In short, culture was the most important foundation for people to maintain a friendly relationship with the environment. It also supports the use of local resource among all kinds of ethnic groups along the basin (Xiao Yun, 2013b).

The Red River mid-reaches in China pass through a nearly 400 kilometer-long valley in the Red River County and in Yuanyang County. Our case study is mainly focused on Yuanyang County. In the valley, Asia's largest terraced rice paddy-fields are scattered over Xinping County, Red River County, and Yuan Yang County. There are millions of hectares of terraced fields along the valley, especially in Yuanyang County and Red River County where the terrace paddy-fields are the most concentrated (Figures 9.1 and 9.2).

The characteristic of the terrace paddy-fields in this area is that of the wet field, whereby people keep the water in the field all year. When visiting the area, we can see a water world in the mountains with the terraced paddy-fields largely covered with water all year (Figure 9.3).

Figure 9.1 A section of Red River. (Photo by Xiao Yun. Zheng).

Figure 9.2 The nature- human scape: forest, village and paddy-fields. (Photo by Xiao Yun. Zheng).

In fact, millions of hectares of terraced fields are based on water resources and vast irrigation systems. Natural rivers alone could not support the water demands of the millions of terraced fields. So, the construction of irrigation canals that follow the natural terrain was the most important method of supporting the terrace agriculture civilization. Subsequently, the construction of irrigation systems accompanies the formation of terraced fields, and the size of the terrace field determines that of the irrigation canals. Therefore, the huge scale and complexity of the irrigation system in the Red River water basin area is astonishing. For example, by 1994, the Red River County, Miangyan County, Liyang County and Jingping County, in the middle of Red River, had already constructed water 12,350 canals, and this number went up to 24,745 in year 1995. The irrigated area is as much as 50,000 ha; the main channels had a flow rate of more than 0.3 cm^2 per second.

Figure 9.3 The terraced paddy-fields in the season without rice. (Photo by Xiao Yun. Zheng).

In the Yuanjiang County of the upper reaches of the river, 79 canals had been constructed by the early twentieth century with irrigated areas of 350 ha. There were also more than 2000 other small canals. By the early 1980s, 2,300 water canals had been constructed in the whole county used for irrigating farmland of 5,000 ha (The Editing Committee, 1993). As canals of an irrigation system, this is not only amazing in terms of number, but also in scale. Some water canals were as long as 10–20 kilometers, while other small water canals were linked by bamboo pipes and functioned as water pipelines for as long as 2–3 kilometers. From small water channels of hundreds of meters to large water canals of dozens of kilometers, built by continuous work over hundreds of years by different ethnic minorities' resident along the Red River, this has formed an enormous irrigation system that supports the local agricultural civilization. Thus, as vital as the arteries of a human body, the irrigation system continuously providing water to the millions of paddy-fields. This made it possible to go into rice paddy agriculture along the valley, providing a source of livelihood to millions of people along the Red River.

9.2 WATER CULTURE AND ITS ROLE IN WATER MANAGEMENT

Water culture is structured systematically at four levels: ideas, institutions, social life, and physical life. These elements solely or interactionally influence water management. In this chapter, the water culture will be surveyed according to the levels of the structure.

9.2.1 The conception of environmental planning and resource utilization

Agricultural development in the terrace field region goes through three stages, namely: shifting cultivation, fixed cultivation, and terrace paddy agriculture. Terrace paddy agriculture is the most advanced form (Xiao Yun, 2005). When planning local agriculture and the living environment, local knowledge of water culture was used. Based on the natural and geographical conditions in the area, the Hani people divided the local natural environment into three functional zones: the forest zone for water resources, the residential zone, and the terrace cultivation zone. In Duo Yi Shu village, Yuanyang County, there is a complete picture of human ecology combining forests, villages, and terraces. The summit of Eastern Guanyinshan Mountain is about 2,930 meters above sea level (asl). Above 1,800 meters asl the mountain was covered in forest – the forest protection zone. The main function of this zone was to conserve water as as natural reservoir, so that it could be taken into the villages below by artificial drainage, and then allowed to flow into the farmland to meet the needs of people and irrigation. This area was also the main source of the river in the east of Yuanyang County. At an altitude of approximately 1,600 to 1,800 meters asl, many villages were distributed on a line, from west to east. These villages include: Da WaZhe, Luo Bu Zhai, Lian Ban Cha Chang, Ai Chun Ha Dan Bu, Ai Chun, Ai Chun Da Yu Tang, Ai Chun A Zhe Ke, Ai Chun Niu Luo Bu, Duo Yi Shu, Hou Zi Zhai, Pu Gao Xin Zhai, Pu Gao Lao Zhai, Huang Cao Ling, and Deng Cun Zhai. Below the villages, the terraced fields extend from 1,500 meters asl to the foot of the valley at 600 meters asl. This landscape is typical of the middle reach of the Red River. The core of this wise planning of the environment was the local people's conception and knowledge of water. People's clever use of local natural feature led to the construction of a triune system consisting of forest, village, and terrace paddy-field.

9.2.2 The conception of water conservation

The local people of different nationalitiies have different conceptions of water which are directly related to the water use, management, and conservation of water. A very important element is that people carried out water conservation as part of their religion. They believe that water is a place where gods live and as such needs to be protected. The local villagers believe that there are spirits in the woods that surround their settlements and so the woods should not be destroyed. Usually, every village has a mountain to be identified as a spiritual site that is to be protected. Towards the end of every year, people worship and sacrifice to the gods of forest and water and pray for safety in the coming year. Therefore, when

New Year Day arrives, people have to sacrifice to the forest to pray to God to bless the people with enough water. Simultaneously, people also cannot enter the forest to hunt, log, get firewood, and graze. Even in some of the core areas of forest, people are banned from entry. People believe God has provided a steady stream of water. Hence if you destroy the forest, then this would offend God, leading to disaster. A typical example is the local Hani people who worship a kind of tree, which local people call the Chinese Water Melon tree, due to a mystical story. This story told that there were three Chinese Water Melon trees behind the mystic-master's house. The trees are water springs and the springs always have abundant water all through the year. Therefore, this story gave rise to a Hani saying that Chinese Water Melon tree is a water spring. This tree was popularly planted in the mystic forest or around the villages and paddy-fields. These trees are carefully protected by the villagers, and people cannot cut them down or even chop its branches. When people wish to rest after work, they always like to sit near this tree. They can only sit close by because they cannot lean on the tree. Hani people like to name their villages after the Chinese Water Melon tree. Indeed, there were more than ten villages named after the tree in Nichu County, for example, including Ni long, Niu bo, Niu wei (Figures 9.4 and 9.5).

Due to the power of God, people cannot cut down the trees. Thus, religion is the most important way of forest protection. In addition to irrigation water supply system, local villages have water supply facilities for daily use, including wells and springs, raceways, etc. Such water facilities are to be protected alongside the daily habits of the people. Additionally, religious practices also played an

Figure 9.4 The houses near the field. (Photo by Xiao Yun Zheng).

Figure 9.5 A woman washing vegetables at the tank. (Photo by Xiao Yun Zheng).

important role. For example, people do not only maintain the wells and clean the springs frequently, the villagers also have to undergo religious rituals which serves as a means of protection. Every year in February, the villagers in Hani village in Yuanyang County offer rituals to the god of the well. During the ritual, the leading member of every household of the village must attend, and people have to kill a pig and a hen to worship the god of the wells. After the ritual, people will repair and clean up the wells and ask the god to constantly take care of the water wells so that they can be enough water and food for them.

All in all, in the terrace area in the Red River basin regardless of which ethnic group or the process of water management, religion is a very important factor. Also, water and people's spiritual world are closely related. Religious ideas are a very important way to strengthen people's understanding of the value of water, to protect the water environment and water facilities, and to ensure the social behavior is water friendly.

9.2.3 The social norms in water management

In addition to religious factors, there are social norms and social practices to manage water in the local society. In each village, there are many canals and each canal has a dedicated person responsible for daily management and drainage repair. In each village, there is a meeting of elders held to discuss issues such as the management of drainage, water distribution, worship of the god of water, etc. Also, there are social norms and regulations regarding the destruction of drainage ditches, people who goes against the rules of using water, water theft, and other acts. In fact, the local canal irrigation system is linked with a complex social network. Some of the canals which transport water from the forest on the mountain to the terraces are 20~30 kilometers in length and have to pass through many villages, and these villages may be even be homes to different nationalitiies. Therefore, it is a complex task to coordinate the benefits and to ensure the co-operation of different ethnic groups and different villages for water management. In fact, there are good local social norms to coordinate the villages or even different nationalities for water management. The cooperation concerning the drainage has brought out various management functions and coordinated different interests relations. The villages in the water source area (top of the mountain) have the main responsibility of the management of forests is to ensure the forest provides an abundant water supply. The villages where drainage flows through are responsible for managing the canal, as well as for coordinating the distribution of water, economic interests, and even offering compensation to the villages of water source areas. In regard to the canals flowing through the villages, close cooperation must be achieved in the drainage basin for effective management and coordination of interests.

The case study in Yuanyang County, Dong Pu canal is suitable for understanding how different villages and ethnic peoples cooperate in the management of the canal. Dong Pu canal originated in the virgin forest, more than 2,000 meters asl behind Dong Pu Da Zhai village. It ended in terraces in Panzhihua township, is 10 kilometers in length, with an irrigated area of about 70 ha. Dong Pu canal was built by Dong Pu Da Zhai village (Hani ethnic group) in the middle Qing dynasty. Its irrigation basin mostly belonged to the villagers of Dong Pu Da Zhai village. Later, many people from other places moved to acquire or rent rice fields along the basin at different periods. Today, the canal passes through Dong Pu Da Zhai village, Wu li Zhai village (Hani ethnic group), Shan Wu Zhai village (Yi ethnic group), Ma Lao He Cun village (Hani ethnic group), Gie Fang Cun village (Hani ethnic group), and Bao Shan Zhai (Yi ethnic group). It is apparent that the canal flows through many villages, which are inhabited by different nationalitiies. Thus, the task of coordinating the management and sharing the benefits of the canal is hard work.

Firstly, a person who is notable in the community was chosen during the yearly leaders meeting as the leader of the canal from the villages of the basin. He is in charge of the management of the whole drainage system and the coordination of village affairs related to the canal, including maintaining the canal, coordinating the responsibility and water use, and sharing the benefits among the villages of the basin. At same time, people have social norms for maintaining the safety of the canal and public benefits; for example, there was a customary law among the villages to restrict people's behaviors on water use. The customary law states that people cannot get water without uniform management or use water over the fixed amount, steal water, destroy irrigation facilities, etc. Also, it stipulates the appropriate sanctions that accrue to defaulters (Figure 9.6).

Secondly, there is a council in charge of coordinating the canal affairs. Every year, the villages along the canal annually sent representatives to discuss matters

Figure 9.6 The complex water diversion channels. (Photo by Xiao Yun Zheng).

relating to the canal, such as water distribution agreements, canal repairs, volunteer worker from the villages, collection of rice for the canal management, etc. Furthermore, the most important topic of the meeting is the evaluation of the canal manager's achievements and his responsibilities over the past year. If people think the manager failed in his responsibilities, a new one will be elected. On the other hand, if the people were satisfied, suggestions will be given for further improvements.

According to the decision of the canal management meeting, the villages along the drainage system must be responsible for the repair of the canal and ditch. Water from the canal is shared into units. A unit is about 15 cm wide broad of about 15 cm of the channel, and villagers usually consider the size of their paddy-fields to decide how many units they need. After they decide, then they will pay rice for it. The standard payment for each unit of water was 15 kg rice per year. The revenue from the villages along Dong Pu canal each year was around 900 kg. It was used to pay the water manager and also as compensation to the village of the water source.

9.2.4 The heritage of water environment and water facilities construction

Terraced paddy-fields represent the grand tangible heritage in the mid-reaches of Red River, as the terrace landscapes do not just consist of agriculture, but are also rich in human meanings such as spiritual elements, social institutions, and social behavior. The terrace landscape, with its physical elements (paddy-field, irrigation system, facilities of water sharing, water-source forests, and techniques) serves people's livelihood needs but also becomes the cultural symbol of the nationalities.

In recent years, these terrace landscapes have become hot spots for tourism and research, and they represent an important resource for local development. Every year, there are hundreds of thousands of tourists who visit the terrace area, bringing income to local people and cultural influence at the same time. Actions to protect the terraces will also be effective in environmental protection.

Taking the case above, the role of water culture is exhibited in the following ways. Firstly, based on their understanding and awareness of water, people planned and reconstructed their living environment: they created irrigation systems, constructed a grand terrace landscape, and developed agricultural civilization dependent on the terraced field. Secondly, the religious factor was an important for water conservation. Local people take their religious worship to the water source and the god of canals to express respect to the god of water, and limit people's behavior toward water. Thirdly, water culture promotes cooperation among people in villages to form social norms for water management. In the drainage basin, people in each village have to cooperate in drainage management and distribution of benefits. As a result, a management group for water affairs in the basin was formed. People also have annual consultation meetings on water management, ritual, distribution of benefits, and the obligations which people must bear. Good collaboration in the drainage basin is the foundation for the effective management of water.

9.3 VALUE AND ROLE: WATER CULTURE IN AN APPLIED PROSPECT

Water culture is not just an important aspect of human beings we inherited from the past, but it also plays a significant role in water management and utilizes it for water sustainable use. Water environmental protection from the past to the present, from an ethnic group to a nation, from the rural to the city, especially its values on present age, still requires more research and education.

Based on the above analysis, the values and role of water culture in general could be reflected as described below.

Water culture is favourable to people in retaining the concept of water. The concept includes people's understanding of the relationship between water and the origin of the world. It entails understanding the importance of water for human livelihood, traditional knowledge, and religious beliefs about water. Since a concept of water first came in to existence, people have cherished and protected water. The maintenance of traditional concepts of water and traditional knowledge of water uses, water protection, reasonable use, and distribution and coordination of water management are very important elements for the sustainable use of water in the contemporary age. For example, religious factors play a positive role in water management in the Red River basin, showing that water culture has a spiritual level.

Water culture can effectively regulate people's behavior to water so as to establish a form of friendly behavior towards water. The case of many nationalities in the Red River basin and other examples in Yunnan shows that there are patterns of social behavior to protect water and water facilities. Local social norms require that people cannot destroy water sources and forests in the water source area. This means no forest logging, no grazing, ensuring drainage and water facilities are not damaged, no washing of dirty clothes near the water facilities, and no taking of water without permission (Xiao Yun, 2008). This also shows that establishing a kind of social behavior which is conducive to the protection of water is very important in the sustainable use of water.

The social norms and social regulations of water management and use have been developed in the long historic process of water management. These social regulations and social norms mainly include two aspects. One aspect is relevant customary behavior, that is the regulations and norms concerned with water management and protection. There are written and unwritten regulations, and social rules regarding water use and water protection in each village. Acts harmful to water or behavior that does not meet the generally accepted rules and regulations on the management of water and use are condemned. Another aspect is that the social norms and regulations have promoted people's cooperation in water management, including the repair of canals, worship of the water gods and the implementation of obligations towards collective water management.

9.4 CONCLUSION

In conclusion, this case study shows that water culture can be observed in real society, and it plays a very important role in water management. Water culture is beneficial for people to keep water values for sustainable use, promote cooperation for water management, and ensure protection of the water environment. It is very important in supporting the local subsistence as well. This study emphasises that water should not be managed without the culture. However, there is a danger that the paddy-field system could collapse. Unfortunately, fast social change and urbanization has emerged outside the valley but has an influence on the valley life which changes the traditional industry at the local level. In recent years, tourism quickly developed in this valley, and the terraced paddy landscape has become a very popular sight, especially now been listed on the UNESCO World Heritage in 2013. These changes are not beneficial to the water culture as the local people are moving away from their traditional subsistence. Therefore, conservation of the water culture in a fast changing society is a major challenge. Despite that, water culture is still a key factor for the local water sustainability. If water culture collapses, it will also lead to the collapse of the local water management and brings unsustainability to the local subsistence. Therefore, maintaining the traditional water culture and building a new culture to meet the current demand and situation of water management is very important for the sustainable livelihoods.

REFERENCES

The Editing Committee. (1993). The County-graphy of Yuanjiang Hani Yi People's Autonomous County. Zhonhua Books Bureau, Beijing, p. 188.

The Editing Committee. (1998). Annals of Yunnan Province – Annals of Water Resources. Yunnan People's Publishing House, Kunming city, p. 321.

Xiao Yun Z. (2005). Ethnic Water Civilization and Irrigation in China Reaches of Red River. Forming Japan, Tokyo, **39**(5), 5–41.

Xiao Yun Z. (2008). Water Culture: A New Way to Understand the ECO-Civilization of the Ethnic People in Yunnan. Yunnan Educational Press House, Kuming, p. 78.

Xiao Yun Z. (2013a). The theory and prospect of water culture. *The Thinking*, **4**, 1–9.

Xiao Yun Z. (2013b). National Culture and Ecological Civilization of Red River Basin. China Book Press, Beijing City, p. 112.

Section 4

The Cultural Role in the Local Societies

Chapter 10

Agricultural water management customs in Japan: Adaptive changes, recent trends, and future issues

Mari Takeda

ABSTRACT

This paper focuses on the customs of community-based irrigation management (CBIM) in Japan, which represent Japanese water culture. There are two aspects to the customs of irrigation canal management: (1) the custom of assigning maintenance work responsibility among stakeholders; and (2) allocation of maintenance workload. These customs have changed over time due to socio-economic changes in rural communities and the physical changes in irrigation. In this sense, adaptive change in water culture at the rule-in-use level may contribute to maintaining CBIM. Depopulation and aging in rural society have a severe influence on CBIM and government policy, which promotes public involvement in agricultural water management. In the future, promotion of public participation may influence water culture and create a sustainable collaborative relationship of irrigation management among stakeholders.

Keywords: agricultural water, collective, irrigation, paddy-field, maintenance, adaptive changes, Japan

10.1 INTRODUCTION

Rice cultivation is a fundamental factor in the history of water resource development and management in Japan. Traditionally, water resource development was driven

Editor: Xiao Yun Zheng
doi: 10.2166/9781789062045_0161

mainly by the expansion of paddy-fields because rice was used as a currency, and this was handed over as a form of tax payment in old Japan. Consequently, the development of paddy-fields had a direct impact on the regional economy. At that time, the water flowing through the paddy-field irrigation canals was also used for daily life upstream and to fill moats of castles downstream. Thus, people naturally recognized the multipurpose nature of water. This meant that agricultural water was recognized as a regional resource, which may imply that local water-related culture may have, more or less, a connection to agricultural water use.

Table 10.1 shows aspects of Japanese culture deeply related to water that are recognized by the Japanese people living in urban areas (Tokyo, Nagoya, Osaka). About 40% of the respondents recognize that the rural landscape of paddy-fields and irrigation canals, with its long history, has been a representation of Japanese water-related culture even though the share of farmers nowadays is only 3% of the Japanese population aged 15 years and above.

Although rice consumption today (57 kg per person) is half of its peak in the 1950s, the current share of agricultural water use is still 67.2% of annual water use (80.9 billion cubic meters in total). This means that agriculture is still the largest consumer of water (Ministry of Land, Infrastructure, Transportation, and Tourism, 2016). Since the amount of water used in dry land irrigation is very limited, irrigation for rice production has been fundamental in national water use.

The sustainability of rice production has been supported by the local custom of community-based irrigation water management. Since this custom can be recognized as representative of Japanese culture (Tabayashi, 1987), it is the focus of this study. In the following sections, the history of irrigation development and

Table 10.1 Aspects of Japanese culture deeply related to water (2016–2017) (% of respondents).

Rank	Aspect	Average	2017	2016
1	Hygienic tap water	55.7	56.1	55.3
2	Rice production (idyllic scenery of paddy-fields)	42.7	43.9	41.5
3	Clear and multifunctional toilet	42.3	45.1	39.5
4	Mountains and forests storing water	39.3	38.7	39.9
5	Bath customs (*Onsen* and public baths)	37.1	37.5	36.7
6	Japanese sake and brewing culture	30.7	32.3	29.1
7	Japanese gardens	25.1	26.1	24.1
8	Japanese-style dishes (raw fish)	23.3	24.8	21.7
9	Japanese paper	19.3	21.7	16.9
10	Ecosystem (World Heritage (Shiretoko, Yakushima))	19.1	21.3	16.9

Source: Water Culture Survey conducted by Mizcan Holdings Co. Ltd.

the institutions for irrigation water management are briefly reviewed. Furthermore, changes in local customs that arise due to changes in socio-economic conditions in rural areas, rice production, as well as the technological development of irrigation facilities are discussed in order to argue for the relationship with social change and cultural change. Finally, future issues are identified through a discussion of recent trends in irrigation water management.

10.2 CUSTOMS AND HISTORY OF PADDY-FIELD IRRIGATION MANAGEMENT IN JAPAN

Japanese paddy-field irrigation management has been characterized by community-based management of common-pooled resources (Hatcho & Matsuno, 2005; Nagata, 1971; Sarker & Ito, 2001; Tamaki *et al.*, 1984). In many cases, the present-day irrigation management system in Japan is established in village communities that were originally formed based on the local units of water use between the late medieval period and the pre-modern Edo period (1603–1867) (Tamaki & Hatate, 1974). In such a community-based resource management system, balancing the efficient allocation of scarce resources and the long-term equity among members is a key point for management of sustainable self-organized common-pooled resources (Hayashi, 2008; Shogenji, 2006; Takeda, 2008). In the context of community-based irrigation management (CBIM), there are two fundamental customs. Firstly, there are local customary rules for water distribution such as '*Koden Yusen*' (older rules, similar to a prior appropriation rule), '*Jyoryu Yusen*' (upstream served rules), and '*bansui*' (rotational irrigation during water shortage periods). Secondly, there is the custom of equal contribution to irrigation facility maintenance (one person from each household in the community should attend to the collective maintenance work, such as dredging and weeding, at the lateral canal level and/or on-farm level).

Traditionally, farmers refrained from opportunistic behaviors, such as free-riding on the maintenance workload of others, because 'there was the credible threat of ostracizing the opportunistic family from participating in the other spheres of the social, political, and economic life of the village' (Aoki, 2001, p. 46). 'The other village families could refuse to cooperate with a shirker by denying him mutual aid and excluding him from participation in social events in the community' (Aoki, 2001, pp.46); in other words, 'community norms can be a stable outcome linking an irrigation domain and a community social exchange domain' (Aoki, 2001, p. 43) from a theoretical point of view. Although this ostracism does not obviously occur at present, villagers know its history. Due to the long-term settlement in villages, the social exchange domain is still important for inhabitants and opportunistic behavior can be avoided.

Shiono *et al.* (2017) describes the history of water management organization in the early modern period in the case of multiple villages depending on a single irrigation scheme; to cover a wider area a united water users' association would be formed (Tamaki, 1982). In the Meiji period (1868–1912), water users' associations were regulated by modern laws to strengthen the government's control over grassroots water use. Although such legal associations were officially made up of landowners, the actual management on the ground was assumed by villagers as a continuation of pre-modern water users' associations (Tamaki, 1984).

In 1949, the Land Improvement Act was enacted to encourage the implementation of large-scale irrigation schemes, and many informal united water users' associations were transformed into modernized legal bodies, for example, Land Improvement Districts (Kobayashi, 2005). Although a Land Improvement District is a group of water users who manage new irrigation facilities installed under the above law, it is also a farmer-based water management organization based on existing local communities.

Thus, customs of water distribution and facility maintenance that have a long history can still be in effect in community-level irrigation management. Recent research has mainly focused on the field of facility management. Since completion of a demographic transition and a shrinking agricultural sector caused by economic growth eventually reduces the growth in water demand, competition for water only happens locally or temporarily. In other words, the scale and frequency of water disputes among farmers or among sectors is decreasing due to economic development. Therefore, water use efficiency in the agricultural sector may not be the most critical issue in a depopulated and developed country.

In contrast, there are several issues in sustainable community-based irrigation facility management, especially at the lateral canal level or on-farm level. Since aging and depopulation are most severe in rural areas, the changes to these socio-economic conditions of the community and individual irrigation users may reduce the number and the quality of contributors to maintenance work,

Figure 10.1 Collective maintenance work for irrigation canal cleaning: (a) March, 1978; (b) August, 2017. Source: (a) Tedorigawa Shichika Yosui Land Improvement District (1982), (b) Yamagata Prefecture.

especially manual labor work to maintain on-farm irrigation facilities, such as dredging and weeding. Therefore, the traditional customs and organization of community-based collective maintenance work may be malfunctioning. According to the Japanese government, the estimated cost of the manual labor was 7.4 million yen, which is about 30% of the total cost of maintenance of the facilities (Ministry of Agriculture, Forestry, and Fisheries). Since the community mobilizes most of this manual labor, CBIM has a significant effect on irrigation maintenance work (see Figure 10.1). Therefore, the custom of the on-farm or village irrigation facility management will be discussed in the next section.

10.3 CHANGES IN THE CUSTOM OF IRRIGATION FACILITY MAINTENANCE

10.3.1 Introduction

As already explained, the custom of collective maintenance work, that is, the equal allocation of maintenance workload per household (in many cases, not only agricultural households but also all households in the community), gradually malfunctioned due to the aging and depopulation of farmers. In addition, this custom has been gradually recognized to be unfair by farmers because of the imbalance in the costs and benefits of the maintenance work: the workload of large-scale farmers is equal to that of small-scale farmers under the traditional allocation, but the large-scale farmers gain greater benefits from the irrigation facilities. This notion of unfairness may be strengthened in accordance with the heterogenization of farmland scale among farmers and may increase the free-riders for maintenance work and result in a further decrease in contributors. The decrease in the number and quality of contributors to maintenance work would affect the efficiency of the irrigation water supply in the whole irrigation system and hinder the promotion of long-life facility management. The key point in addressing this issue is to analyze the possibility of adaptive change of two aspects in light of these socio-economic changes: (1) the custom of assigning maintenance work responsibility among stakeholders and (2) the custom of maintenance workload allocation among them.

10.3.2 Adaptive change of the custom of maintenance work responsibility assignment

10.3.2.1 Background

Recent studies on irrigation maintenance work have tended to focus on factors promoting public involvement in maintenance work because of the difficulty of sustainable maintenance by farmers and non-farming farmland owners alone, as well as the growing awareness of the various benefits of irrigation facilities (Aizaki *et al.*, 2006; Ashida *et al.*, 2005; Honda, 2011; Yamamoto *et al.*, 2010). In 2007, the Japanese government launched a nation-wide policy that encouraged not only farmers but also various kinds of stakeholder, such as non-farming residents, local firms, not-for-profit organizations (NPOs) and schools, to participate in maintenance work because these stakeholders can also benefit from the irrigation facilities. The irrigation facilities are important elements of the rural landscape, they provide diverse environments to enhance biodiversity, and they are used for firefighting and snow melting in some areas. These various applications of the irrigation facilities demonstrate their 'multifunctionality,' which is one rationale for the necessity of public involvement in the maintenance work. However, since this recent public involvement is a policy-driven movement, this section focuses on the custom of community-based irrigation facility management before 2007, and analyzes how the custom of assignment of maintenance work responsibility was changed or unchanged in accordance with the changes in the socio-economic conditions of the rural community. Policy-driven public involvement in irrigation facility management will be discussed in section 4.

10.3.2.2 Types of stakeholder

Stakeholders for the maintenance work are divided into three categories: (1) cultivators (who work at least a certain area of farmland but also rent farmland); (2) non-farming farmland owners; and (3) non-farming residents (who own neither farmland nor farm, that is, they have no connection to agriculture). All three types of stakeholder can be in charge of the maintenance work in Japan. In other words, the person in charge of the maintenance work is decided at the community level since the maintenance work is usually organized by the community. Since cultivators are the fundamental beneficiaries of the irrigation facilities, it may seem obvious that the cultivators should be responsible for the maintenance work.

However, if farmland holding is the critical factor in choosing the assignment of responsibility, not only cultivators but also non-farming farmland owners can be in charge. The cultivators and non-farming farmland owners are referred to as 'landowners.' Furthermore, if all residents are recognized as the beneficiaries of irrigation facilities regardless of farming, not only landowners but also non-farming farmland owners can be in charge of maintenance work. These stakeholders are known as 'residents.' Table 10.2 shows the stakeholders and the possible responsibility assignment patterns.

10.3.2.3 History and principles of responsibility assignment

Before the choice of responsibility assignment is explained, it is necessary to briefly explain its history in maintenance work. Traditionally, the responsibility was assigned to 'residents' because all residents used the irrigation canals for their everyday life activities and in rice production. After the Second World War, land reform was implemented by the Japanese government under General

Table 10.2 Stakeholders and assignment of responsibility for maintenance work.

Stakeholders	Type of Responsibility Assignment		
Cultivators[a]	Cultivators	Landowners	Residents
Non-farming farmland owners	–		
Non-farming residents	–	–	

[a]Since cultivators usually have their own farmland in Japan, they are known as owner-cultivators.

Headquarters Command in 1947. This reform created homogeneous same-scale owner-cultivators (Shogenji, 2006). Since all cultivators owned their own farmland and almost all residents were involved in agriculture in rural areas at that time, there were only a small number of non-farming farmland owners and non-farming residents in rural areas, if any, at that time. Therefore, the responsibility for the maintenance work was assigned to the 'cultivators'. This implies that in reality, all residents had responsibility.

In 1961, the agricultural structural policy was launched with the intention of expanding farming scales in order to resolve the economic disparities between farmers and industrial laborers. Therefore, some of the owner-cultivators gave up farming and started working in other sectors, renting out their farmland to other farmers who were willing to expand their farming scale. Non-farming farmland owners emerged in this process, and the right and responsibility of non-farming farmland owners for the irrigation facilities became an issue.

The behavioral principle of communities that organize maintenance work is to minimize the transaction cost of organizing it. Therefore, if a community that previously assigned the maintenance responsibility to 'landowners' cannot continue the maintenance work because of, for example, landowner depopulation, the community may change the responsibility assignment only when it is able to achieve a consensus to do so. In some communities, non-farming residents have a reason to participate in the maintenance work (they benefit from the irrigation canals that are partially used for storm drainage, etc.). Thus, this may reduce the transaction costs of achieving a consensus and mobilizing non-farming residents for the maintenance work. However, even though non-farming residents have such a reason to contribute, there may be cases where communities have difficulty achieving a consensus because non-farming residents are reluctant to participate when they derive only a small amount of indirect benefit from the canals or have difficulty with labor work due to lack of experience. In such cases, the community may keep the original responsibility pattern as long as the maintenance work can continue. On the other hand, the community might give up organizing the maintenance work and the collective maintenance work would be abandoned. In this sense, the community's choice of responsibility assignment includes the option of abandoning the organization of collective maintenance. Thus, the choice of responsibility assignment can be determined by various

Table 10.3 CBIM and assignment of responsibility for maintenance work.

Year	Number of Rural Hamlets	Share of Hamlets[a]	Share of Hamlets CBIM	Responsibility Assignment			Share of Hamlets N-CIM
				R	C	I	
1980	133,638	90.9%	69.9%	–	–	–	30.1%
1990	139,413	91.8%	75.6%	42.4%	56.6%	1.0%	24.4%
2000	135,163	90.8%	78.4%	39.9%	59.6%	0.5%	21.6%
2010	139,176	90.4%	73.2%	–	–	–	26.8%
2015	138,256	90.3%	78.5%	–	–	–	21.5%

R: Residents; C: Cultivators; O: Outsourcing; N-CIM: Not organized by community. Source: Census of Agriculture and Forestry in Japan.

factors, such as the scale of the benefit from the facilities, experience and knowledge of agriculture, and the number and age of the stakeholders. Table 10.3 shows statistics on the CBIM and responsibility assignment.

The relevant factors for the choice of responsibility assignment can be divided into two categories: physical factors and socio-economic factors. Physical factors, such as the location of the irrigation facilities, may offer some benefits. In cases where the irrigation facilities are open (rather than pipes buried underground) and run through not only paddy-fields but also residential areas, irrigation water may be used for miscellaneous purposes (rainfall drainage, washing vegetables, watering the garden, etc.) or may provide a good rural landscape. In addition, there are situations where the irrigation facilities have a special multifunctional benefit (biotopes, etc.) or can be used for additional purposes (firefighting and snow melting, etc.). In these cases, not only cultivators but also non-farming farmland owners and non-farming residents may be considered as beneficiaries. However, due to the introduction of the pipeline system, which makes maintenance work unnecessary, communities may decide to abandon collective maintenance work and even CBIM.

Socio-economic factors include the number of cultivators in a community, the proportion of non-faming farmland owners to cultivators, the age profile of the cultivators, and so on. Although the Land Improvement District laws determined that cultivators had rights and responsibilities for the land improvement project, there was no room for practical issues to emerge through which to assign the rights and responsibilities for irrigation facilities' management because non-farming landowners did not exist at that time. In addition, the responsibility for the maintenance work was usually practical and regarded as a local matter; thus, the law does not formally or clearly defines them.

In some cases, socio-economic factors indicate that the cultivators can sustainably carry out the maintenance work; concretely, where the number of cultivators is large enough, cultivators' aging is not serious and the accumulation of the bonding social capital among the cultivators is significant. In these situations, the community may rationally decide to assign the responsibility to the 'cultivators'. This is because the community can maintain the facilities and save the transaction costs of requiring non-responsible persons to join in the maintenance work. However, if this is not the case, the community may look for an opportunity to change the responsibility assignment. For example, if the community has a sufficient number of homogenous landowners, the community may decide to assign the responsibility to 'landowners.' On the other hand, severe depopulation of the 'landowners' may again lead the community to make a decision to abandon the collective maintenance work or even CBIM itself.

10.3.2.4 Empirical results

Table 10.4 shows the summary statistics of communities' responsibility assignment as surveyed by the national agricultural census in Japan.

These were the only data that showed responsibility assignments for the whole country, and one of the limitations is that the surveys that gathered the data were carried out only in 1990 and 2000. In addition, there was no direct question about responsibility assignment to 'landowners.' We could access only data on responsibility assignment to 'cultivators' and 'residents.' Since non-farming landowners are in a middle position, between cultivators and non-farming residents, it is not easy to estimate the possibility of responsibility assignment to 'landowners' using these data. However, this is a good period to examine the data because it was not influenced by the national policy implemented after 2000 to provide financial support for self-governing community-based collective maintenance.

Takeda (2015) conducted panel-data analysis using the above data to identify the factors in changing the responsibility assignment customs. Since responsibility assignment is a discrete choice, a fixed-effects model is not applicable. Therefore, a correlated random-effects (CRE) model was used, which is the middle ground between a random-effects model and a fixed-effects model (Wooldridge, 2010). Takeda's analysis showed that homogeneity in the community might induce the assignment of responsibility to 'residents.' By contrast, a negative coefficient of the aging variable implies that aging may reduce the possibility of assigning responsibility to 'residents.' In addition, residents' recognition of their direct benefit from the facility (draining miscellaneous wastewater into the canal) may encourage communities to assign responsibility to 'residents,' although the facility's development may change this recognition.

Based on the results of the empirical analysis, we conclude that communities adaptively change the custom of responsibility assignment for the irrigation facilities in order to suit the changes in the socio-economic factors of irrigation users and the physical conditions of the irrigation facilities. This adaptation may prolong CBIM. However, the depopulation of farmers in a

Table 10.4 Summary statistics of responsibility assignment change of 1990 and 2000.

Numbers of Agricultural Hamlets		**2000**					**Total in 1990**
		Residents	**Cultivators**	**Outsourcing**	**Not Organized by Community**	**No Irrigation Canals**	
1990	Residents	19,653	13,059	122	4,647	2,034	39,515
	Cultivators	11,222	30,009	214	8,435	2,257	52,137
	Outsourcing	233	392	33	219	62	939
	Not organized by community	5,025	9,747	91	9,116	2,550	26,529
	No irrigation canals	864	1,021	13	1,339	4,716	7,953
Total in 2000		36,997	54,228	473	23,756	11,619	127,073

Source: Census of Agriculture and Forestry in Japan.

community and aging can have a critical effect on CBIM itself. In addition, adaptive changes in the custom of responsibility assignment cannot provide a solution in this case.

10.3.3 Adaptive changes in the custom of maintenance workload allocation

10.3.3.1 History and principles of maintenance workload allocation

Another important aspect of the custom of maintenance work is the rule of allocating workload among the responsible stakeholders. Historically, one person from each responsible household (cultivators, landowners, or residents) should participate in the one-day or half-day collective maintenance work (such as canal dredging) (Honda, 2013; Nagata, 1971; Nagahama, 2003; Shogenji, 2006; Takeda, 2008). This equal share rule of the maintenance workload among responsible households may, when the responsible households are homogeneous, follow the 'benefit principle.' According to this, the cost of maintenance work should be borne in proportion to the benefit derived from the irrigation facilities. Since the benefit principle is one of the key principles of long-enduring institutions for common-pooled resources, including irrigation (Ostrom, 1990; Sarker & Ito, 2001), this principle can be an important factor in retaining CBIM. However, heterogenization of the responsible households may undermine the benefit principle because the responsible households' benefits from irrigation facilities might be different. The community might need to change the equal share rule to retain the benefit principle.

The common way to adjust to heterogenization is monetary adjustment, usually in the form of *monetary compensation*, *outsourcing*, or *payments*. In regards to *monetary compensation*, since the traditional rule (the equal share rule of maintenance work) under community-based irrigation is fundamental, the responsible households are expected to participate. On the other hand, *monetary compensation* is recognized as a penalty for absence. For example, in cases where the community assigns the responsibility for the maintenance work to 'cultivators' and some large-scale cultivators cannot participate in the maintenance work for the canals located in their cultivated plots, they pay a sum of money as compensation so they can be exempted from participating in the maintenance work on some occasions. In cases where the responsibility is assigned to 'landowners' and non-farming landowners have to go to work on the day that collective maintenance work is done, they provide *monetary compensation* and are not allowed to join in the maintenance work. Thus, benefit principles can be maintained between participants and absentees. According to the case study conducted by Honda (2013), monetary compensation can offer an incentive to avoid free-riding by the non-farming residents because it is expected as a penalty for absence.

Monetary adjustments also include methods that are more market oriented: *outsourcing* (outsourcing of maintenance work) and *payments* (payments to the responsible villagers engaged in the maintenance work). Outsourcing does not require collective maintenance work and/or CBIM. The laborers hired by the local authority or formal water user organizations (such as Land Improvement Districts) engage in the maintenance work. By contrast, the use of payments premises CBIM because the community usually has an accounting system for local irrigation facility management and has a payment arrangement with the participants. Despite the difference between outsourcing and payments, they offer further ways to retain the beneficiary rule because the funds tend to be collected from the cultivators or the landowners as an area-based fee. Thus, monetary adjustment provides a way for the community to retain the benefit principles under the heterogenization of the households responsible for irrigation management.

10.3.3.2 Empirical results

Table 10.5 shows the results of the empirical analysis of the monetary adjustment choices in the sampled Japanese irrigation project areas surveyed by the Ministry of Agriculture, Forestry, and Fisheries in (2017). The dependent variable is the choice of monetary adjustment (if a community chooses monetary adjustment (monetary compensation, outsourcing, or payments), select 1; otherwise, 0), and it is estimated by logistic regression (Takeda, 2008).

In the empirical analysis, the samples are divided into two subgroups according to the responsibility assignment types: 'cultivator' and 'landowner.' In cases where the responsibility for the maintenance work is assigned to cultivators, the heterogeneity of the cultivated area among the cultivators can be an issue for the beneficiary rule because the benefit derived from the irrigation facilities depends on the cultivators' area. Therefore, the independent variable 'Index of heterogeneity of cultivated area' is included for the subgroup 'cultivator.' Since this variable is statistically significant at the 10% significance level (p-value is 0.063), the heterogeneity of the cultivation area among the cultivators may induce the choice of monetary adjustment. Hence, the benefit principles may be maintained under heterogenization among the cultivators.

In addition, the independent variable 'Share of area cultivated beyond border of hamlet farmer lives in' is positive and statistically significant. This may suggest that the farmers who cultivate beyond their hamlet's border are another factor in heterogeneity. They may be inclined not to clearly recognize the rule designated for the maintenance work for canals along the plots located beyond their community and may become free-riders on the maintenance work there. Thus, monetary adjustment may be required as the share of the area cultivated beyond the border of the hamlet where the farmer lives increases.

Table 10.5 Main results of choice of maintenance work rule (unit: canal).

Dependent Variable (Labor with Monetary Adjustment = 1, Only Labor = 0)			
Type of Responsibility Assignment		**Cultivator**	**Landowner**
Independent variable			
Cultivators' heterogeneity	Index of heterogeneity of cultivating area among cultivators*	1.121 (1.86)*	–
	Number of farm households per hamlet	0.048 (2.65)***	–
	Square of number of farm households per hamlet	−0.0002 −1.42	–
	Aging rate of farmers	−8.771 (4.43)***	–
	Number of households per hamlet	–	−0.002 −0.39
	Square of number of households per hamlet	–	−0.00002 −1.02
Community members characteristics	Share of non-farming farmland owners	–	0.265 −0.2
	Share of farm households	–	16.436 (4.32)***
	Square of share of farm households	–	−13.999 (4.30)***
	Aging rate	–	−10.298 (2.48)**
	Existence of new residents zone (1: Exists, 0: Does not exist)	–	4.978 (6.75)***
	Share of area cultivated beyond border of hamlet farmer lives in	1.756 (5.45)***	4.935 (5.57)***
	CBIM dummy (1: Yes, 0: No)	−0.937 (2.74)***	−0.586 −1.28
Institutional factors	Collective maintenance work dummy (1: Yes, 0: No)	4.297 (10.67)***	4.938 (3.48)***

(*Continued*)

Table 10.5 Main results of choice of maintenance work rule (unit: canal) (*Continued*).

Dependent Variable (Labor with Monetary Adjustment = 1, Only Labor = 0)			
Type of Responsibility Assignment		**Cultivator**	**Landowner**
	Number of agricultural organizations	−0.71	−0.089
		(6.32)***	−0.81
	Community farming organization contributing to maintenance work dummy	0.428 −1.03	−0.195 −0.47
	Number of non-agricultural organizations contributing to maintenance work	–	−0.255 (3.06)***
	Number of activities not related to agricultural production in a year	−0.25 (8.31)***	−0.014 −0.44
	Number of hamlet meetings in a year	−0.012	0.033
		−1.08	(2.83)***
Constant		−5.468	−6.314
		(4.03)***	(2.25)**
Sample size (Unit: irrigation ditch)		1,260	1,235
LR chi2(32)		1043.91	561.99
Prob >chi2		0	0
Pseudo R2		0.6223	0.4908

Note 1: Index of heterogeneity of cultivating area among cultivators was calculated by the sum of large-scale registered farmer's paddy-fields over total paddy-fields.
Note 2: Both estimation models include regional dummies (urban, middle, mountainous, Japanese regional categories), economic factors (share of abandoned paddy-fields etc.), and physical factors (canal conditions etc.) that are not shown in the table.
Note 3: Number of valid responses is 228 sites with 2,495 irrigation ditches divided into 'cultivator' (1,260 samples) and 'landowner' types (1,235 samples).
Note 4: ***, **, * statistical significance at 10%, 5% and 1% respectively.
Source: Takeda (2008).

On the other hand, in cases where the responsibility of the maintenance work is assigned to 'landowners,' the heterogeneity of landowners' relationship to agriculture can be an issue. This means that if all the landowners are farmers, monetary adjustment may not be necessary because they can all benefit from the maintenance work. By contrast, if only some of the landowners are farmers, some non-farming landowners may feel that they do not benefit from the maintenance work. In Table 10.5, such heterogeneity among landowners was

Figure 10.2 Landscape and biodiversity enhancement in irrigation canal. Source: Yamagata Prefecture.

captured by the independent variables 'share of farm households' and 'square share of farm households.' Since these coefficients show an inverted U-shaped relationship between the choice of monetary adjustment and the share of farm households, it may suggest that as the heterogeneity of landowners increases in terms of their relation to agriculture, the choice of monetary adjustment increases also.

Therefore, these results may suggest that changes in the traditional rules for maintenance workload allocation may occur to retain the benefit principles under the heterogenization of the persons responsible for the maintenance work, although endogeneity issues remain in the statistical analysis. These adaptive changes may permit the continuation of CBIM itself.

10.4 RECENT TRENDS AND FUTURE ISSUES

In section 3, adaptive changes to the customs of the CBIM at rule-in-use level were discussed. The empirical analysis may suggest that the custom of responsibility

Figure 10.3 Public involvement in fish release in irrigation canals and cleaning a farm pond. Source: Yamagata Prefecture.

Table 10.6 Public involvement in irrigation canal maintenance work.

Region	Total Number of Rural Hamlets	Number of Rural Hamlets with Irrigation	Number of Rural Hamlets with CBIM	Share of CBIM with Urban Residents	Share of CBIM with NPO, Schools and Firms
Hokkaido	7,081	5,161	3,228	1.36%	0.96%
Tohoku	17,432	16,406	13,966	2.31%	0.54%
Hokuriku	11,050	10,734	9,210	1.43%	4.94%
North Kanto	9,043	8,342	4,544	2.46%	0.57%
South Kanto	8,908	7,381	5,069	3.35%	0.49%
Tohsan	6,341	5,917	4,998	1.32%	0.08%
Tohkai	11,613	10,598	8,136	3.99%	0.68%
Kinki	10,796	10,292	8,956	2.77%	0.27%
Chugoku	19,663	18,149	14,580	2.23%	0.39%
Sanin	5,718	5,116	4,261	1.71%	0.31%
Sanyo	13,945	13,033	10,319	2.44%	0.43%
Shikoku	11,027	9,781	7,898	3.67%	0.27%
Kyusyu	24,552	22,021	17,409	1.76%	0.39%
Okinawa	750	414	218	0.92%	0.92%
Total	138,256	125,196	98,212	2.38%	0.86%

Source: Census of Agriculture and Forestry in 2015.

assignment and the custom of maintenance workload allocation may change adaptively in line with socio-economic changes, which may contribute to maintaining the CBIM system. If the local customs are considered as a representation of the water culture, cultural change may contribute to sustainable self-governing CBIM.

However, continuing depopulation and aging in rural areas would make it more difficult to sustain CBIM. CBIM is also recognized as a water culture or even as Japanese culture itself (Toyama, 1980). In this sense, trends in CBIM are very important when discussing water culture. As mentioned in the previous section, the Japanese government launched a national policy to support CBIM, and recent trends have been influenced by this policy. This policy has implemented direct payments to communities or alternative local entities to support community-based agricultural resource management since 2000 in less favored areas and since 2007 in all agricultural promotion areas. The latter policy requires public involvement because public understanding and cooperation will be necessary for agricultural resource management in the near future due to farmer depopulation and aging. In addition, since non-farmers also benefit from the multifunctionality of irrigation facilities (through, for example, enhancement of biodiversity, landscape conservation, and land conservation, see Figures 10.2 and 10.3), collaborative work on maintenance is rational in terms of the cost–benefit relationship. Public involvement promoted by the national policies is not intended to be the same as the traditional communities' responsible assignment of 'residents.' This is because this collaboration is established based on consensus among the stakeholders.

Although these policies may have a significant impact on maintaining CBIM (Komiyama & Itoh, 2017; Takeda, 2018), the share of public involvement in the whole country is not high as at present (see Table 10.6). Furthermore, innovations in irrigation management using pipelines and information and communications technology may offer other solutions for irrigation management (Iida *et al.*, 2016; Takeda, 2017). These technological solutions may make collective maintenance work unnecessary, as in the case of drinking water infrastructure. Strengthening public involvement in collective maintenance work and promoting pipelines and information and communications technology may be solutions that undermine the need for CBIM. Thus, this means that Japanese rural society may face a turning point in terms of its community-based resource management system. How water culture influences the transition in irrigation water management systems, and vice versa, may be topics for future research.

REFERENCES

Aizaki H., Tsuchiya Y., Kondo T. and Osanami F. (2006). The conditions related to non-farmer participation in agricultural canal maintenance activities: the case of

Watari Town, Miyagi Prefecture, [in Japanese]. *Japanese Journal of Farm Management*, **44**(2), 1–11.
Aoki M. (2001). Toward a Comparative Institutional Analysis. MIT Press, Cambridge, USA.
Ashida T., Yagi H. and Kunimitsu Y. (2005). Maintenance of Irrigation canals by a local community [in Japanese]. Proceedings of Annual Meeting of the Japanese Society of Irrigation, Drainage and Rural Engineering, Gifu City, Gifu Prefecture, Japan, pp. 206–207.
Hatcho N. and Matsuno Y. (2005). The participatory approach of managing water in Japan. In: A History of Water Issues, R. Coopy, J.H. Fahlbusch, N. Hatcho and L.r Jansky (eds.), United Nations University, Tokyo.
Hayashi K. (2008). Use rights assignment in the commons and the fairness problem arising therefrom [in Japanese]. *Journal of Rural Planning Association*, **26**(4), 416–426.
Honda Y. (2011). Factors affecting non-farmers' participation in the maintenance of irrigation and drainage canals: focusing on styles of irrigation water management and canal maintenance [(in Japanese]. *Journal of Rural Planning*, **30**(1), 74–82.
Honda Y. (2013). Chiikishigen Hozensyutai to shite no Shuraku - Hinoka, Shinjyumin sanka ni yoru Saihen wo Mezashite (Community as a Local Resource Conservation Entity: Toward a Reorganization by Non-farmers' and New Residents' Participation). Association of Agriculture and Forestry Statistics, Japan.
Iida T., Kimura M. and Kubo N. (2016). Value evaluation of the irrigation information service system for rice cultivating famers in Japan. Paper presented at the Second World Irrigation Forum, 6–8 November 2016, Chiang Mai, Thailand.
Kobayashi H. (2005). Japanese water management systems from an economic perspective: the agricultural sector. Paper presented at the OECD Workshop on Agriculture and Water: Sustainability, Markets and Policies, 14–18 November 2005, South Australia.
Komiyama M. and Itoh J. (2017). Policy evaluation of the farmland, water and environmental conservation improvement scheme: A case study of Shiga Prefecture. *Journal of Rural Problems*, **53**(2), 72–83
Ministry of Agriculture, Forestry and Fisheries. (2017). Maintenance of Land Improvement Facilities. http://www.maff.go.jp/j/nousin/mizu/agwater_antei/a_totikai/ (accessed 19 December 2017).
Ministry of Land, Infrastructure, Transport, and Tourism. (2016). Water Resources in Japan (in Japanese). Retrieved on 19 December, 2017, from http://www.mlit.go.jp/common/001049556.pdf (accessed 19 December 2017).
Nagahama K. (2003). Chiiki Shigenkanri no Shutai Keisei -Shuraku Shinsei he no Joken wo Saguru (Forming Local Resource Management Entities: Searching for the Conditions for Community Revitalization). Nihon Keizai Hyouronsha, Ltd., Tokyo, Japan.
Nagata K. (1971). Nihon Nogyo no Suiri Kozo (The Structure of Irrigation in Japanese Agriculture). Iwanami Shoten, Tokyo.
Ostrom E. (1990). Governing the Commons: The Evolution of Institutions for Collective Action. Cambridge University Press, Cambridge, UK.
Sarker A. and Ito T. (2001). Design principles in long-enduring institutions of Japanese irrigation common-pool resources. *Agricultural Water Management*, **48**(2), 89–102.
Shiono M., Ikegami K. and Tsuruta T. (2017). Climate resilience of collective water management in rural Japan. *Journal of Asian Rural Studies*, **1**(2), 162–171

Shogenji S. (2006). Gendai Nihon no Nosei Kaikaku (Policy Reform in Japanese Agro-food Sectors). University of Tokyo Press, Tokyo.

Tabayashi A. (1987). Irrigation systems in Japan. *Geographical Review of Japan*, **60**(1), 41–65.

Takeda M. (2008). Institutional Analysis on Water Resource Allocation and Management: Focusing on Agricultural Sector (in Japanese). Doctoral thesis, University of Tokyo.

Takeda M. (2015). Factors to influence a community's choice for on-farm irrigation facility management responsibility assignment to all households within the community in Japan: panel data analysis using rural community cards. *Japanese Journal of Agricultural Economics*, **87**(3), 262–272.

Takeda M. (2017). Recent innovations in agricultural water management in Japan. Program book of the First Asian International Water Week: Asian Solutions for Water, 20–23 September 2017, Gyeongju, Republic of Korea.

Takeda M. (2018). Rural Policy and Rural Community: Program Evaluation Using Difference-in-Difference Estimations (in Japanese). General Analysis of the Results of the 2015 Census of Agriculture and Forestry. Association of Agriculture and Forestry Statistics, Tokyo, Japan.

Tamaki A. (1982). Nihon no Shakai System (The Social System in Japan). Nousan Gyoson Bunka Kyokai, Tokyo.

Tamaki A. and Hatate I. (1974). Hudo: Daichi to Ningenno Rekishi (Hudo: History of its Land and People). Heibon Press, Tokyo.

Tamaki A., Hatate I. and Imamura N. (1984). Suiri no Shakai Kozo (The Social Structure of Water Use). United Nations University. Tokyo.

Tedorigawa Shichika Yosui Land Improvement District. (1982). Tedorigawa Shichika Yosui Shi (Historical Record of Shichika Irrigation water in Tedori river). Tedorigawa Shickika Yosui Land Improvement District, Ishikawa.

Toyama K. (1980). Mizu no Bunka-shi (Cultural History of Water). Chuko-Shinsyo, Tokyo.

Wooldridge J.M. (2010). Econometric Analysis of Cross Section and Panel Data, second edition, MIT Press.

Yamamoto T., Kita T. and Nagasawa T. (2010). Consideration of factors to promote participation in maintenance works of irrigation and drainage facilities (in Japanese). *Journal of Rural Planning*, **30**(1), 74–82.

Chapter 11

Development of water conservancy irrigation and local social reconstruction: A case study in Putian, Fujian Province, China

Longjiang Du

ABSTRACT

The Putian Plain is essentially the land created from a coastal swamp with the support of a huge hydraulic system, which means that the relationship between man and water has been decisive for the rise and fall of Putian. With a sophisticated water system, the Putian people created artificial plains from the coastal swamp, by a process similar to land reclamation in the Netherlands. The water system is used to control the water level on the plain, drain seawater, introduce freshwater, and develop the land. Under such a man-and-water relationship, focus is given on the relationship between the hydraulic development and social organization in Putian Plain. In the field investigation, it was found that the development of Putian Plain was based on the construction of awater irrigation system, and the management and distribution of water resources was realized through civil organizations, which created a unique folk ritual alliance and traditional cultural etiquette that helped to reconstruct the social organization and interest space in Putian Plain. It also helped to form a stable social organization structure and achieve a balance between cooperation and conflicts. The farmland water systems formed during the history of Putian not only reflect the development of agricultural productivity, but also reflect the reconstruction of social organization in the production process. It is, however, a typical case study of the relationship between water resources management and social organization.

Keywords: irrigation, Putian, water conservancy, dam

Editor: Xiao Yun Zheng
doi: 10.2166/9781789062045_0181

11.1 INTRODUCTION

11.1.1 Water conservancy development in Putian

The Putian Plain, where the study area is located, is the third largest plain in Fujian Province, China. It reaches to Xinghua Bay in the east, to Jiuhua Mountain in the west, to the hills of Hengtang in the south, and to the southern part of Nanshan in the north. It covers an area of 464 km^2, accounting for about 23.5% of the land area in Putian County (Shaohong and Zeng Congsheng, 1991). It is the most densely populated area and the main grain and fruit producing area for Putian City. The largest river in this area, the Mulanbei, twists and turns in the middle of the plain and empties into Xinghua Bay via the Sanjiang Estuary. The Putian Plain is formed by both terraced sediments in the rivers and artificial reclamation. It is only 3 to 7 meters above sea level and there are about 800 villages. The plain in the north of Mulanxi is referred to as the 'Beiyang Plain' and the part in the south is called the 'Nanyang Plain'. Thus, Putian Plain is also known as Nanbeiyang Plain. The plain is criss-crossed by a dense network of channels. Topographically, the plain tilts towards Mulanbei in the north, west, and south (Chen Yangchi, 1974) (Figure 11.1).

11.1.2 Early development

Putian Plain used to be a bay on the southeastern coast of China. Later, the deposited sediment gradually formed a swamp that was repeatedly affected by the tides.

Figure 11.1 Satellite view of putian plain in Fujian.

Therefore, to grow crops in this locality, first of all, a seawall had to be built to defend the land against the tides. At the same time, it was necessary to build a water reservoir and use the freshwater to wash the salt out of the soil and irrigate crops. As early as the Tang Dynasty (AD 627–649), people built small-scale water storage pits and ponds in the north and south of the plain[1], with an irrigation area of 1,200 ha. Later, Yanshoubei was built and the irrigated area increased by 400 ha. Several water storage and irrigation projects have been built since then. The Beiyang water reservoir was first developed in the Tang Dynasty (Yishhou). The watercourse was extended from Yanshou to Dutang and then to the sea, reaching to Lupu in the south, to Ninghai in the southeast, and to Xijiang in the northeast. The water was stored to irrigate fields and the whole area was called Beiyang. During the Tang Dynasty, Wu Xing promoted reclamation of the sea, built a dike in Dufu which stopped the water flow and led it southward into Shatangban, and later named it Yanshoubei (Chen Kang *et al.*, 2004). According to records, the tides in Beiyang during the Sui and Tang Dynasties rose as far Shihua Bridge and Chengqing Mountain, and all the rivers in the Beiyang flowed to the sea from the villages of Dutang and Qi'ao. After Wu Xing reclaimed the land, he built an embankment in the Dutang areas, and then set up sluice gates in Qixi and Daban, the seawater receded[2]. According to records at that time, these water conservancy irrigation projects were concentrated on the transition zone between hills and plains, with no record of construction in the middle of the plain. This shows that at the time, the Putian Plain was still not fully reclaimed and its middle part was still a salt marsh that was not suitable for development (Ge Jianxiong, 2000).

In the past, when people went northward to Fuzhou, they usually pass through the mountainous area after leaving Putian City by the north gate. For fear of the frequent attacks by tigers living in that area, people were forced to change their route to lower-lying places[3]. Based on this record, it can be seen that the Putian Plain was basically below sea level during the Sui and Tang Dynasties, and some independent water conservancy projects only appeared at the junction of mountains and plains.

11.1.3 Construction after the song dynasty

Hydraulic engineering development in Putian flourished during the Song Dynasty (AD 960–1279). At this time, a long embankment was built on the Beiyang Plain. The ponds in the embankment were used to provide freshwater for irrigation. There are more than ten such embankments (Wu Songdi, 1995)[4]. The

[1] 'New Tang Book' Geography Volume, Putian County.

[2] Chen Chiyang, 'Puyang Water Resources' Volume 3.

[3] Zhou Wei, 'Hongzhi Xinghua History' Volume 52.

[4] Liu Kezhuang, Song Dynasty, 'Renewed Three Steps'.

Beiyang Plain was reclaimed by encirclement, and the Nanyang Plain became farmland after the construction of the Mulanyu Water System. In general, development of Putian Plain progressed gradually from north to south and from west to east. The original reclamation area was only near the county town, then advanced to the middle of the plain, and then continued extending eastward (Li Xiaocong, 1986).

After a thousand years of continuous development, the Putian Plain water system became very complex and composed of large and small river canals, scattered over the plains of the Nanyang and Beiyang, with a total length of 309.5 km and an irrigation area of 10,867 ha. The length of the Nanyang and Beiyang embankment is 87.48 km. Along the coastal embankment, there are 17 tidal gates with 57 distributaries, 82 culverts, 24 control gates, and 91 spur dykes. The water conservancy system in Beiyang is a networked water system that has gradually merged with several independent water conservancy systems after many years of development. In contrast, the Nanyang Plain water system has a herringbone pattern, which was highly planned and was built within a short time. The backbone of the Nanyang Plain water system is the Mulanbei water irrigation system. The main canal of Huilan bridge carries water from Mulan to Nanyang, and then divides into two branches, one entering the Qingjiang River in the northeast direction, and one turning to the Huangshi in the southeast (Liao Biqi and Wang Dajing, 1879). The entire Nanyang water system is generally divided into upper, middle and lower sections, where the main channel is divided into branch channels and then subdivided into sublateral canals, according to a hierarchical structure. Overall, the development of the Beiyang water system was organic without a unified and clear overall planning; it shows a network-like pattern, which implies traces of its millennium history. The Nanyang water conservancy irrigation system, represented by Mulanbei, is well planned, uniformly implemented and effective in the short term. It provided for explosive growth in the social landscape and productivity of the Nanyang Plain. It is a masterpiece of the construction of water conservancy systems in Putian and underlies the social structure and cultural traits of Putian today (Liang Kejia, 2003).

Nanyang was widely developed after Li Hong, who had been posted to Putian as a public servant, constructed the dykes in the Song Dynasty (Local Records Compilation Committee of Fujian Province, 1990). He learned the lessons of the failure of his predecessors, and chose a base where the height difference between the flood peak of the Mulan Mountain and the sea level was the least. After the success of the building, the water was drained by channels and seven big ditches and countless small ditches were excavated. They re-established four sluice gates on the coast. After the completion of the project, many ponds such as Hengtang were abandoned and converted to farmland, leaving only the Guoqing Pond. The north part of the Small Turtle Island and the east of the Great Turtle Island were reclaimed producing a land area of more than 200 ha. After Li Hong excavated the channels and built the sluice gates to make a large-scale water conservancy

Figure 11.2 Nanyang plain in putian.

project in Nanyang, the appearance of Nanyang also changed. In the Yuan Dynasty, people continued the operations and converted Guoqing Pond to farmland, which covered 2 hectares. People also consolidated the land near the sea in Chengkou and Dongzhang, and therefore expanded the area of Nanyang cultivated land (Tuo Tuo, 1977). In the Ming Dynasty, the reclamation of land around the coast continued (Figure 11.2 and Figure 11.3).

Figure 11.3 Beiyang plain in putian.

11.1.4 Hydraulic development method

The water conservancy development in Putian Plain is of two types, lake reclamation and coastal reclamation, forming two irrigation systems of diversion and accumulation. This means of water control according to local conditions is effective and is the creation of local people. Lake reclamation is mostly in the lower reaches of the river, where lakes and rivers crisscross, and the reclaimed land is extremely vulnerable to flooding (Xie Ruming, 2008). This kind of water conservancy construction involves dredging and is the most popular near the cities and in the lowlands.. For the highlands above the lake, the method of using terrain to draw water is adopted. Lake reclamation, both in highlands and lowlands, established irrigation by using canals.

Coastal reclamation was the most typical development method for the coastal people in Fujian during the Song Dynasty. From the original endless coast, a huge sea levee of ten or twenty miles was built, and tall sluice gates were built to control the entry and exit of the tide. A large area of land was formed inside the levee. The saline soil must be flushed by freshwater, so water storage and 'resist the tide' projects and the salt-flushing systems are necessary water conservancy facilities. Along the reclaimed saline land, people also built dams and ditches according to local conditions. Putian is near the sea, where the elevations are low and the springs are small. These water conservancy projects not only mark the scale of the project from small to large, but also show people's water conservancy planning progress from scattered and blind temporary measures to a highly systematic level (Figure 11.4).

Figure 11.4 Starting dykes in mulanbei in putian.

11.1.5 Water conservancy and social organization reconstruction in putian

The social organization form of Putian water conservancy can be roughly divided into two types: 'government-run' and 'privately-run'. The early development of Putian Plain was done mainly by the government. The large-scale farmland water conservancy in Putian Plain, represented by the Mulanbei Irrigation District, covers hundreds of thousands of hectares. The prosperity or waste of the farmland can affect the overall situation and, generally, it has been cultivated and managed by the government during different dynasties. The area that benefits from the prosperity this farmland is far greater than the irrigated area. Large-scale farmland water conservancy facilities often have various functions such as water storage, irrigation, and drainage, which are difficult to be organized and managed by the private sector.

The farmland water conservancy system in Putian Plain was officially repaired during the Song and Yuan Dynasties, and detailed regulation were formulated. In the Ming Dynasty, 'the sea people reclaimed new field outside the embankment, and gradually opened wider and wider, naming these places the first dam, the second dam, the third dam, and even the seventh dam. More of the dams were located, lower than where the land was, where there was less salt and more freshwater. So it was called the dam-land. Or besides the official dykes, the local people privately built steep gates on the outer embankment, and built culverts to the sea'[5]. This kind of water conservancy, which was privately owned by the 'sea people', can only be attached to the government-run water conservancy system due to the lack of independent water sources, and thus it will be subject to the government. There is also an interdependent relationship between the various civilian water conservancies in the Hekou Plain. The above-mentioned dams, sharing the same water source, actually belong to the same water conservancy system and are relatively independent and interdependent, so that various contradictions are inevitable. It is worth noting that this kind of privately run farmland water conservancy project not only needed the help of the government, but also often needed to be hosted by a few highly respected monks or country gentlemen for its effective building and management.

11.1.6 Change from government run to privately run

The first signs of the change from government-run to privately run were already seen in the early Ming Dynasty when the government handed over the responsibility for water conservancy to the local landed gentry. More importantly, since the Ming Dynasty, the local government has had no special funds for the construction of

---

[5]Zhu System, Ming Dynasty 'On the Nanyang Water Conservancy with Wu Taishou', quoted from Daoguang 'Fujian Tongzhi' Volume 34, 'Water Conservancy'. Putian County.

water conservancy. Therefore, even if the officials come forward to organize the construction of water conservancy, the funds were still from local people. Before the middle of the Ming Dynasty, local governments raised funds for the construction of water conservancy, mainly in the form of a levy. The government adopted a similar approach to the 'duty-by-field system', which completely transformed the official responsibility into a civil service. In the local literature on the coast of Fujian in the Ming and Qing Dynasties, although there are many records of 'official constructions' in water conservancy, in fact, it is nothing more than an organizational form of 'official promotion and civil construction' or 'construction with the help of the government.' These are all transitional forms from government-run to privately run. Around the middle of Ming Dynasty, the management style of government-run water conservancy also showed significant changes. The government handed over the responsibility of management to township organizations, and legislated for this.

In the past, the shortcomings of the government-run water conservancy in the Ming and Qing Dynasties were often attributed to the corruption and incompetence of local bureaucrats. In fact, bureaucratic corruption is not a unique phenomenon in the Ming and Qing dynasties. Therefore, it cannot be used to explain the shift and recession of the government-run water conservancy. The reason why the government-run water conservancy in the Ming and Qing Dynasties went in to decline was that the local government lacked financial power. In the Song Dynasty, although the financial power of local bureaucrats was also very limited, the financial resources specially used to build water conservancy were quite abundant. At the same time, there were also official lands for government water conservancy. In contrast, in the Ming and Qing Dynasties, the local governments did not have the funds for water conservancy, neither did they have official land for this purpose. In the Ming Dynasty, some large-scale water conservancy projects were still available for funding. In the Qing Dynasty, there was no such policy. Since government-run water conservancy can only be funded by the people, the government-run water conservancy is bound to change to privately run.

The power of the government alone is not enough to exercise effective macro control over the water conservancy cause. Therefore, it is necessary to rely on the guarantee of the clan rules and social rules and even rely on the gods. Due to this, in the farmland water conservancy process along the coast of Fujian in the Ming and Qing Dynasties, the role of the feudal government was weakened, and the power of the township organization grew stronger. In a certain sense, the transition from the government-run to privately run reflects the development of township organizations and landed gentry powers. Since the private sector has not formed an authority strong enough to replace the local government, the transition from government-run to privately run cannot be fully realized. In this transition, with the emergence of civic leaders, the social structure of Putian is bound to change (Figure 11.5).

Figure 11.5 The carved stone showing the water management of putian in the qing dynasty.

11.1.7 Civil autonomy and the water conservancy community

The aforementioned water conservancy project is a farmland water conservancy system based on water storage and irrigation. It is characterized by interception along the rivers. The farmland water conservancy project is generally maintained and managed by the farmers who benefited from the project. These farmers jointly undertook the responsibility of repairing and managing dams, and formed a fairly strict rural social organization. According to records, Shandeba was built by 'all the families in the whole country' of Wenxianli, and there were thirty-three persons in thirteen surnames which were registered as 'dike households'. This way of 'dike households registration was different from the general tenants because it was 'ordered by the dikes' and 'by the fields', that is: 'the field was repaired among the households according to their number. Each household gets six acres of land and builds its own irrigation system.' If these water conservancy projects were built by the local Liu family, for example, their related rights will be shared only by the members of the Liu clan and no outsiders will be allowed to participate. Family rules and township agreements are the 'written laws' of township organizations. Therefore, the farmland water conservancy system within the villages has a legal effect once the family rules and township agreements are formed. Every clan organization was responsible for the maintenance and management of the farmland profit-raising business organized by themselves, and its related rights were also shared only in the tribes. The costs associated with this were generally shared among the landowners within the clan. However, for some water conservancy projects organized by clan organizations, the beneficiaries were not limited to clan groups. Clan organizations could also collect water rents from all involved fields in the name of maintenance. For this reason, there was always a monopoly of clan organizations on water resources.

In areas where the clan power was insufficient to monopolize the water source, the water conservancy project was jointly organized by local residents. This type of organization based on geographical connections may include several villages, a thousand families or many individuals, and in terms of its organizational principles, it generally has the characteristics of a joint stock operation. As mentioned above, the 'Dike households' of Shandeba in Xianyou County consists of thirteen surnames and thirty-three persons. They may not have repaired all the farmland and the land size of each household may not be the same. Therefore, in order to align the rights and obligations of the beneficiaries, it is necessary to adopt the principle of 'repairing the dam according to the land shared'. Here, the 'dike household' is not considered as an individual, but represents the established water-covered field and the established equity of the dam. This form of joint-stock operation also applies to water conservancy projects co-organized by several clan or several villages. Generally speaking, township organizations that jointly organize water conservancy projects have are similar water distribution regulations. Within the township organization, the rights and obligations of the water conservancy cause are directly combined, thus effectively mobilizing the enthusiasm of the people to build water conservancy, and better realizing the distribution and utilization of water resources. However, it should also be pointed out that such township organizations are exclusive, which inevitably lead to monopoly on water resources, and to various water disputes and conflicts. Under these circumstances, the organizational form of farmland water conservancy depends not only on natural conditions, but also on the social power system. Therefore, in different historical periods, similar farmland water conservancy projects may adopt different organizational forms, which necessarily reflect the changes in rural social structure.

11.1.8 Water conservancy system and 'seven realms' alliance

The farmland water conservancy project in Putian is generally maintained and managed by 'waterfield households'. The above-mentioned 'waterfield households' jointly undertake the responsibility of repairing and managing the water conservancy system, forming a fairly strict rural social organization, which is generally built by local clan. Also, its related rights are shared by the tribes and no outsiders are allowed to join in. This type of hydraulic production has gradually evolved into a ritual alliance with the Putian characteristics of 'seven realms.'

The so-called ‘realm’ refers to the territory of a ‘society’, and the original meaning of seven realms is an alliance of seven societies. A society is a ritual unit of the Lijia system in the Ming Dynasty, generally referred to as a ‘Li Society’, or a grassroots management system. Later there were many changes. Almost every village or family has its own society. By the middle of the Ming Dynasty, the Li Society gradually developed folk temples. In the early stage, social sacrifice, as a national sacrifice, was completed in the open air, and later evolved into a private sacrifice in a folk temple. But its organizational structure is in one continuous line.

According to field study statistics, there are 3.2 families in each village in Putian Plain, and an average of 4.8 villages form a ‘seven-realms’ ritual alliance. There is an average of s 3.6 temples per village. Some villages have as many as 18 temples, each temple housing an average of 4 god statues, with a maximum of 35 gods, an average of 14.5 gods per village. Every year they celebrate the birthday of the gods and other god-related festivals together. Beyond these ‘seven realms’, there are some larger ritual alliances, usually with common temples and ceremonies, sometimes covering the entire water system. Most of these ritual alliances were formed during the Ming and Qing Dynasties. The earliest dated back to the early Ming Dynasty, while the latest dated back to the late Qing and the Republic of China (Figure 11.6 and 11.7).

The ‘seven-realms’ were originally formed in the cultural spaces such as the water conservancy community. Their essential role is to control water resources

Figure 11.6 Sangong Temple in baihu, putian.

Figure 11.7 Xituan, typical food for ceremonies in alliance celebrations in putian.

and land, and it is also an important guarantee for productivity and wealth. The villages divided into seven realms have a common point of interest, and the ties connecting them are the water source and the land irrigated by the water system. Although in the Beiyang Plain this phenomenon is not as obvious as in Nanyang, its division mechanism is the same (Figure 11.8 and Figure 11.9).

Why is this ritual alliance formed? What is its relationship with the water system? In other words, what kind of historical and logical connection exists between the water conservancy system and the social organization? Based on my opinion, the foundation of the 'Seven-realms' ceremonial alliance of the Putian comes from

Figure 11.8 God sacrifice ceremony in yuanxiao festival in putian.

Figure 11.9 Seven-realms Parade signboard in lantern festival.

the water irrigation system, that is, several water sub-irrigation areas within the overall irrigation system. In the empirical research view, the two have obvious coupling in space, that is, the spatial scope of the ‘seven-realms’ is mainly the irrigation area with the main watersheds as the core. From the perspective of social organization, the rights and obligations of the water conservancy are directly combined, thus effectively mobilizing the enthusiasm of the people to build a water conservancy, and better realizing the distribution and utilization of water resources. At the same time, such township organizations are exclusive, resulting in a monopoly on water resources and causing various water disputes and conflicts. In this situation, it is necessary to adjust the contradictions between various ethnic groups and to realize the macro-systematic control of the water conservancy caused by relying on the political power or turning to the gods. Failure to follow the correct water conservancy public policy to rationally resolve water resources contradictions can sometimes cause great damage to local social and economic conditions, and even bloodshed, which is not uncommon in the history of Putian. Thus, further research should be carried out on this issue (Figure 11.10).

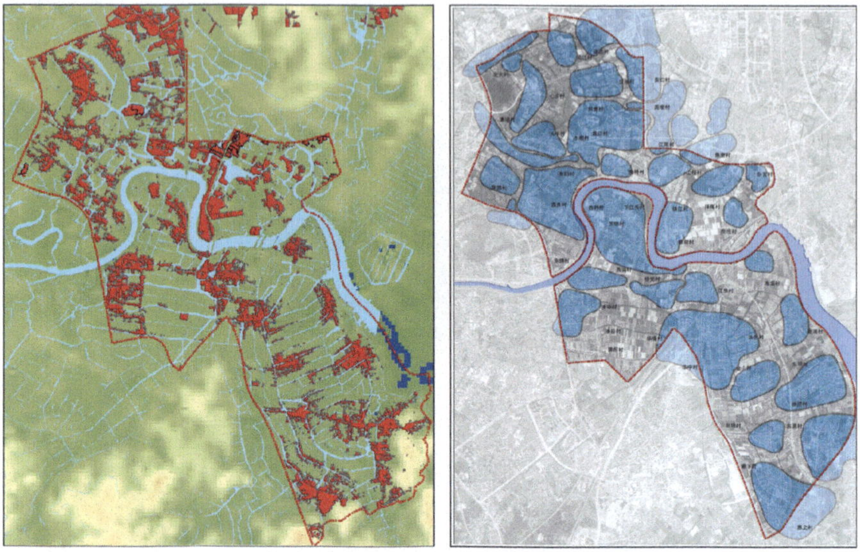

Figure 11.10 Distribution of some natural villages and irrigation canals in nanyang and beiyang: corresponding seven-realms distribution.

11.2 CONCLUSION

Through the field investigation, it was found that the god system in Putian is highly developed. There are more than 2,200 kinds of gods among which more than 1,700 kinds are worshipped. The ceremonial alliance established on this system is also very distinctive, forming a complicated civil society organization. In the empirical investigation, it is found that in areas where water resources are scarce and people-and-land relationship is delicate, the system of gods and social organizations are relatively more developed, such as the 'Lian' system in North China and the 'Seven-realms' ritual alliance in Putian. For such intriguing phenomenon, we try to give an explanation that a very important role of Putian's god system and ritual alliances is the social connection between the people and water. In the distribution and management of water resources by civil organizations, interpersonal cooperation and struggle need to be consolidated and resolved respectively by the god system and ritual alliances, and it is through this process that the local society is re-structured. People's belief in gods and enthusiasm for ritual alliances are derived from the practical need for group identity and public resource determination. Through the ceremonial alliances and the worshiping gods, people clearly define their own needs and the distribution of water resources. From the early intuitive 'dike households' to the gradually evolved 'seven-realms' ritual alliances, the social functions of these organizations have matured and expanded, with its coverage extending from the initial settlement near the irrigation channel to the entire Putian community, achieving the development from water irrigation to local social reconstruction.

REFERENCES

Chen Yangchi (1974). Hydrualic History of Putian [M]. Chengwen Publishing Co., Ltd, Taipei.

Chen Kang, Zhong Wen and Zhang Yushi (2004). Classic Yongle Dadian Local Histories Collection: Volume I [M]. Zhonghua Book Company, Beijing, pp. 1116–1122.

Ge Jianxiong (2000). Chinese Population History: Volume III [M]. Fudan University Press, Shanghai, p. 500.

Li Xiaocong (1986). On the cause of the decadence of the water conservancy project in ancient China [J]. *Chinese Agricultural History*, (3), 26–37.

Liang Kejia (2003). Sanshan History[M]. Chen Shuzhen, Collated and Annotated. Fangzhi Publishing Hous, Beijing, 512.

Liao Biqi and Wang Dajing (1879). History of Putian County, Xinghua Prefecture [Z].

Local Records Compilation Committee of Fujian Province (1990). General History of Bamin [M]. Fujian People's Publishing House, Fuzhou.

Tuo Tuo (1977). Song History [M]. Zhonghua Book Company, Shanghai, 1977, 2209–2210.

Wang Shaohong and Zeng Congsheng (1991). The formation of the Xinghua plain in Fujian and the paleogeographic environment [J]. *Journal of Geographical Sciences*, **46**(3), 341–342.

Wu Songdi (1995). Fujian population study in the song dynasty [J]. *Chinese History Research*, (2), 50–56.

Xie Ruming (2008). A brief History of the Development of Putian [M]. Xiamen University Press, Xiamen, **21**.

Zhou Wei and Huang Zhongzhao (2007). In: Re-issue Xinghua History [M]. Jinyao Cai (ed.), Fujian People's Publishing House, Fuzhou.

Chapter 12

The role of community in behaviors towards water in the red river delta and the central highlands of Vietnam

Nguyễn Chí Bền

ABSTRACT

In Vietnam, the region of Red River Delta is the primary living space of the Viet ethnic group; whereas, the Central Highlands region is currently the home of 11 ethnic minority groups. Water is a valuable natural resource essential to these communities' existence, development and creation. Thus, throughout their long history, these communities have developed a specific attitude towards water: they need, treasure and fear water at the same time. Inhabitants of both the Red River Delta and the Central Highlands worship water gods and they consider these gods supreme beings in the pantheon. The development of the cult of water in the two regions has been attributed to the role of associated communities. However, due to the differences in geographical, geo-historical and geo-cultural positions and trajectories between the Red River Delta and the Central Highlands, the role of the community in the emergence and development of the cult of water in these two regions are also different. This chapter, therefore, will examine the role of the community in the cult of water, which is manifested through: the role of priests, the role of villages and other subjects and objects associated with the cult of water in the two regions. The result of this study provides an overview of culture–religion history developed in different cultural spaces of Vietnam.

Key words: Northern Delta, Central Highlands, community role, cult of water, traditional festivals

Editor: Xiao Yun Zheng
doi: 10.2166/9781789062045_0197

12.1 INTRODUCTION

Vietnam, as described by the late professor Trần Quốc Vượng (1934–2005) and his colleagues in a university textbook entitled *Basis of Vietnamese Culture* (Cơ sở văn hóa Việt Nam) (Trần, 2018), is constituted of six cultural regions. As a contributor to this textbook, I endorse Professor Tran Quoc Vuong's way of dividing Vietnam in this way. The Red River Delta and the Central Highlands are two among these distinct cultural regions. Differences in geo-cultural and geo-economic positions of the two regions have led to dissimilarities in their inhabitants' attitude towards water, the natural entity familiar to all human beings.

12.2 THE COMMUNITY OF THE VIỆT IN THE RED RIVER DELTA AND THEIR WATER GODS

The Red River Delta, also known as the Northern Delta, is a delta spanning from latitude 21°34′N (Lập Thạch district, Vĩnh Phúc province) to the alluvial plain at 19°5′N and from longitude 105°7′E (Ba Vì district, Hanoi city) to 107°7′E (Cát Bà island, Hải Phòng city). The region covers about 14,860 km^2 and occupies 4.5% of the whole area of Vietnam. Within the region, there are two major cities, Hà Nội and Hải Phòng, and other provinces, including Bắc Ninh, Hà Nam, Hải Dương, Hưng Yên, Nam Định, Thái Bình, Vĩnh Phúc and Ninh Bình. In 2016, the Red River Delta had a population of almost 21 million, accounting for 22% of the total population of Vietnam. On average, there were 1,413 people living within a square kilometer. The majority population of the Red River Delta are the Kinh (Việt), and several small groups of Mường people live in Ba Vì district (Hà Nội) and Nho Quan district (Ninh Bình). Therefore, it can be said that the Red River Delta is the land of the Kinh. According to Pierre Gourou: 'a village does not appear in the landscape as a collection of houses, but a block of trees. It is surrounded by a bamboo fence with dense treetops forming a solid rampart, and the shaky bamboo branches bring a green and solemn scene to the village.'

As a key element of the landscape, the village also plays a leading role in the spiritual and social life of the peasants. Peasants are not separate individuals nor citizens of a commune, who only participate remotely in the life of the commune, just like residents of the French countryside; on the contrary, the religious, political and social life of the Vietnamese village is very dynamic, and every peasant participates sincerely and enthusiastically in it with an ambition to take on the role that gets more important day by day, according to Gourou (1936, p. 216). Each village, as such, included a number of families undertaking wetland rice cultivation and living together in a defined space. The whole agricultural land, rivers, lakes and ponds were owned by the village. The village allocated its fields to families to use in compliance with rules practiced and inherited through generations. The most common crop plant chosen by peasants of the Red River Delta was rice which was seen almost everywhere in the region. In 1936, Pierre

Gourou published important data stating that 'total area of Bắc Ninh province is 110,367 hectares, of which land not used for cultivation (housing, rivers, mounds, etc) occupies an area of 22,073 hectares, that is, nearly 20% of the area of the entire delta; rice fields occupy 82,543 hectares, and the land dedicated to other crops rather than rice accounts for 4,965 hectares, that is, about 4.5% of the province's total area' (Gourou, 1936, p. 363). Also, Gourou himself further outlined in a footnote that 'if the percentage of non-rice land in Bắc Ninh is applied to the whole delta, we have the following results: 68,000 hectares of non-rice land and 1,132,000 hectares of rice fields' (Gourou, 1936, p. 363).

Hence, rice was the plant closely associated with the Red River Delta. Back then, the element that most concerned the peasants must have been water. Pierre Gourou, when writing *The Peasants of the Tonkin Delta* in 1935–1936, dedicated the third chapter to the examination of this element: Water as part of material environment. To Pierre Gourou, rice cultivation required a great amount of water. And water was an element that dominated every aspect of the life of peasants in the Tonkin Delta: it could make them rich or poor, gave them enough food or made them suffer starvation, and it could bring them wealthy lives or force them to leave their homeland.

In dealing with water and rivers, the Việt in the Red River Delta have built up a dyke system. It is not the only way to deal with water; however, it was the one chosen by the Việt in the Red River Delta. This behavior reflects the 'adaptability' of the Việt, an attribute I mentioned in my 2008 book, where I borrowed an idea from Professor Cao Xuân Huy to describe the Vietnamese people as 'smooth, flexible and fluid like water… the adaptability is preeminence and a survival secret of our people (Nguyễn, 2018, p. 920). Moreover, once placing the Red River in scale of space and time, we cannot see it as it is today. Rather, the river has brought many changes to the Northern Delta, especially since the Việt have decided to cope with its ferocity by building a dyke system. A consequence of such treatment for the aggressive Red River, however, was frequent dyke breaches before the 1970s, leading to the emergence of sunken areas along the Red River, such as Hưng Yên and Hà Nam provinces.

For wetland rice peasants, water is an object of worship. They have always asked Buddha and the gods for water and rain for their agriculture after a long-lasting drought, as in the following example from a folk song:

I pray to God for rain
to take water to drink
to have paddy fields to plough.

There are different ways to pray for water. Firstly, the worshiped statues are carried from the center of the place of worship to the outside, as happens for example in the festival of Tứ Pháp (four powers) in Thuận Thành district, Bắc Ninh province. Local residents have a procession of Pháp Vân (Cloud Goddess), Pháp Vũ (Rain Goddess), Pháp Lôi (Thunder Goddess), and Pháp Điện

(Lightning Goddess). They recite a prayer: '*Three goddesses visit the festival of Un pagoda, strong wind and rain to make easy earning*. Another way is to bring the statue out of the center of the place of worship to dry in the Sun, such as at the festival of Bối Khê pagoda (Thanh Oai district, HàNội city).

However, water is also a subject to be conquered. Many traditional festivals have been held as a way to express the gratitude towards ancient heroes who protected communities from flooding. Festivals which worship Sơn Tinh (Mountain God) and Thủy Tinh (Water God) in Ba Vì, Hà Tây are manifestations of attempts made by wetland rice peasants to conquer water and control floods. Behind the legendary imagery of Lê Lợi, who returned the sacred sword at Hoàn Kiếm Lake, there is not only a desire for a peaceful country but also an image of a cultural hero who is capable of controlling water. On the other hand, for the Việt in the Red River Delta, water is a natural entity that is both gentle and aggressive, therefore it has been sacralized into a deity: the water god. Moreover, people in the Red River Delta express many different behaviors towards water.

Firstly, water is an environment where traditional festivals take place. The location of the Viet's traditional festivals is sacred and secular. Many of these festivals take place in a sacred space which is a river or at sea. For instance, the whale worship festivals of coastal residents from Ha Tinh to Ca Mau provinces take place at river mouths, coastal areas or on the sea. At these festivals, the most typical ceremonial activities are boat racing and swimming contests which take place, for example, at the festival of Bạch Hạc temple (in the town of Việt Trì, Phú Thọ province), the festival of Chèm temple (Tây Hồ district, Hà Nội), the traditional festival in Khê Hồi village (Thường Tín district, Hà Tây province) and the Đồng Xâm festival (Kiến Xương district, Thái Bình province). Secondly, water is an environment where sacrificial offerings are absorbed. Villagers, after their ceremonies, usually float the offerings down the river or on the sea as a way to express their respect to worshiped deities. For instance, during the festival of Bình Đà village (Thanh Oai district, Hà Nội), local residents bring Vía cakes to the river, crush them into pieces and throw them into the river.

Water is also an entity representing aspirations for innovation. It is the starting point of change. Việt people use water for washing ceremonies (lễ mộc dục) – a ritual essential to the process of festivals. People believe that water in the middle of river is pure, and that it is the best to clean worshiped statues with this water. This ritual can be seen in many traditional festivals, such as the festival of Bà Tấm temple (Gia Lâm, Hà Nội), the festival of Hóa Dạ Trạch temple (Khoái Châu, Hưng Yên), and the festival of Cuông temple (Nghệ An).

Water is also subject to sacralization. It is transformed into holy figures representing a community's beliefs and aspirations. Animals like snakes, dragons and whales are sacralized and worshiped in a number of traditional festivals. Many festivals of the Việt are centered around the snake, including ones dedicated to the worship of Mother Goddesses, in which the snake appears under the name of Ông Lốt. Beside the snake, other animals worshipped by the Viet are all related to the water environment. Water is the point of departure as well as the point of return for many animals worshiped in the Việt's traditional festivals. Coastal villages from Hà Tĩnh province in southern Vietnam adopt a cult dedicated to whales, creatures living in the open sea. To fishermen, the sea is a crucial resource of income essential to their lives, but it also contains many dangers. For this reason, fishermen believe that whales are their guardian gods and they will rescue them whenever they have to face dangers posed by the sea.

People from one generation to the next have sacralized many supernatural entities into historical–cultural figures that are important to the Viet's spiritual life. Noticeably, before being sacralized and historicized, many among these figures were entities associated with water. Results from an investigation into more than 94 hagiographies of 94 villages in the Red River Delta showed that 25 villages' tutelary gods were mysteriously born as a result of the combination between a woman and a snake or a dragon. For instance, the figure of Linh Lang, worshiped in festivals near Thăng Long imperial city and Lệ Mật village (Gia Lâm, Hà Nội), has his origin in the river. Other examples can be seen in the cases of gods named Trương Hống and Trương Hát who are worshiped by residents living along the Cầu River (on the border between Bắc Ninh and Bắc Giang provinces). Both of them are derived from snakes. As peasants living by wet rice cultivation or fishing, water is a vital element for Việt people. Thus, when people worship water spirits or historical figures – central elements of traditional festivals – they are actually worshipping water.

Because they live in a delta and have to deal with the Red River, the role of their community in shaping the relation between human beings and water deities is of significance to Việt people. When speaking about the village community of the Việt in the Red River Delta, it is necessary to highlight the role of village conventions. 'Every village has its own convention and a given village convention often includes specific content which cannot be found in other conventions of other villages. Moreover, each convention has an aggregation of its own, that is, it is possible to mention a number of issues or aspects that will not be mentioned in other village conventions' (Tran, 1984). Researchers agree that the conventions have many different names, such as village customs, regulations, rules, or agreements.

In Vietnamese history, village conventions appeared as early as the 15th century. Until the 16th and 17th centuries, they were improved. In the 18th century, village conventions became very popular and were conceptualized as Confucian-inspired regulations applied to given villages. When characterizing village conventions and customs, in particular, and village autonomous mechanisms, in general, scholars usually use an idom stating that 'the rules of the King are secondary to the customs of the village.' However, villages could only enjoy their autonomy to a certain extent, while the central state still sought to strengthen its controlling power.

The conventions paid significant attention to the behavior of village members towards water. Interesting examples can be seen in conventions of some villages in Thái Bình province. In 1924, the convention of Đại Hữu commune, Tân Định canton, Tiền Hải district, Thái Bình province, Article 3 mandated that: 'Those who cause damage to the field edge must pay a fine … and have to rebuild it.' Article 4 in the same convention stated that: 'Anyone who waters the field without permission and does fishing that leads to the leak of farming water must be punished … and forced to bail out water for the affected landlord' (Nguyễn, 2000, p. 140). Article 20 in the convention of Noi Trang commune, Hà Lý canton, Duyên Hà district, Thái Bình province, mandated that: 'The duty of the head of village guards and members of village associations is to patrol inside the village and outside in the field to urge tax collection, to look after planted crops, to repair roads and bridges, to ask younger guards to close and open water inlet sluices in order to appropriately provide water to farming fields, and to oversee village farming fields to make sure that no one lets their ducks go inside this area' (Nguyễn, 2000, p. 146) Article 7 in the convention of Hoàng Nông village, Duyên Hà district, Thái Bình province, mandated that: 'The head of village guards is responsible for pouring and draining water whenever it is necessary to do so. He must ask permission from the village administrative board before conducting his duties. Younger guards will immediately arrest those who open the water sluice and make the field dry up. Village administrative board will punish these people' (Nguyễn, 2000, p. 153). Every year, in addition to the festival to worship the tutelary gods, villagers also hold several public rituals, such as the new rice ceremony in October and a ceremony to pray for rain (Nguyen, 2000, p. 255) Thus, in the Red River Delta, the community attaches special importance to preserving water resources, organizing activities to express their beliefs in water gods and developing written agreements to protect water resources Figures 12.1 to 12.4.

12.3 ETHNIC MINORITY GROUPS IN THE CENTRAL HIGHLANDS AND THE WATER GENIE

Meanwhile, the indigenous ethnic minorities in the Vietnam Central Highlands have different behaviours. The Central Highlands comprises many highlands which are adjacent to each other and from north to south include Kontum, KonPlong, KonHa Nung, Pleiku, M'Drak, Buon Ma Thuot, M'Nong, Lam Vien and Di Linh. Their altitudes vary: Lam Vien is the highest at about 1500 m, Di Linh reaches 900–1000 m, and Pleiku about 800 m, whereas Kontum and Buonma Thuot Highlands are only 500 m high. The Central Highlands with bazan (red basalt) soil includes five provinces: Kom Tum, Gia Lai, Dak Lak, Dak Nong and Lam Dong. The region borders Quang Nam to the north, Quang Ngai, Binh Dinh, Phu Yen, Khanh Hoa, Ninh Thuan and Binh Thuan to the east, Dong Nai and Binh Phuoc to the south, and Attapeu (of the Lao People's Democratic

Figure 12.1 Cau river, Bac Ninh Province. (Photo by Nguyen Chi Ben).

Republic) and Ratanakiri and Mondukiri (of the Kingdom of Cambodia) to the west, covering an area of 54,000 km^2. As the Central Highlands has no borders with the sea and is rather close to the equator, it has two distinct weather seasons a year: the rainy season from April to October, and the dry season from November to April of the following year Figure 12.5.

The indigenous ethnic minority groups in the Central Highlands of Vietnam comprise the Jorai, Ede and Churu who speak languages of the Malayo Polynesian linguistic family and the Bahna, Cohor, So dang, M'nong, Ma,

Figure 12.2 Community on the river used water for a joinery ceremony. (Photo by Nguyen Chi Ben).

Figure 12.3 The monk Thich Tam Dinh (abbot at the Xuân Nương temple, village Phú Ninh, commune Phương Định, district Trực Ninh, province Nam Định) worshiping the water god on the river. (Photo by Nguyen Chi Ben).

Gie-Trieng, R'mam and Brau who speak languages of the Mon Khmer linguistic family. These ethnic communities live in villages, with different names depending on the community and ethnic minority group. For example, the Jorai and Churu call their village *plei*, the Bahna calls it *plơi*, the Co ho *bon*, and the Ede *buon*. The ethnic minorities in the Central Highlands 'whatever group they

Figure 12.4 The community takes sacred water to the place of worshiping Xuan Nuong. (Photo by Văn Huynh).

Figure 12.5 Waterfalls in the Central Highlands. (By permission of the Viet Nam Institute for Culture and Arts Studies).

belong to have settled in an autonomous or independent village and live on upland rice farming… For cultivation, they have to settle in villages which are complete populated and smallest units and each occupies an area with distinct borders that cannot be trampled upon by other villages, if not, conflicts would occur' (Nguyen, 1996, p. 521). The economic activities of these indigenous ethnic minority groups are different in their behavior toward water sources. The Bahna, Brau, R'mam and Gie-Trieng practice upland rice farming (dry rice fields), whereas the Churu and Co Hor practice wet rice cultivation. Among the tribes that live on upland rice farming, however, some groups of people are wet rice cultivators. The Ede, for example, live on upland rice farming, but the group of the Ede Bih practice wet rice cultivation; the So Dang people live mainly on upland rice farming, but the M'nong people lives on wet rice cultivation. Regardless of their rice cultivation modes upland rice farming or wet rice farming, all these indigenous ethnic minority groups have strong attachment to the bazan soil highlands, thus water is very important for them. In their beliefs, these indigenous groups consider this natural phenomenon sacred and call it the Water Genie. The worship of river watering places is a sacred ritual for them. For example, for the Jorai people, the Water Genie (or *Yang Ia*) is very sacred, but also close to them,and provides them with pure, freshwater sources Figure 12.6.

The ceremony of worshiping the river watering place is very important for Jorai people and is held anually after the harvest. The oldest village inhabitant, who is the master of the ceremony prays for the Water Genie to come to the villages and to provide them with pure streams, brimful with water. He is the first person to take water in a dried gourd and pour it into a wine jar and then he takes the first sip of wine from the jar through a bamboo pipe. At the end of the ceremony, women in

Figure 12.6 The village elder performs a ritual to collect sacred water. (By permission of Viet Nam Institute for Culture and Arts Studies).

the village carry water home in their dried gourds. Talking about the behaviors of the Jorai people in relation to water, mention should be made of Pơtao Puih and Pơtao Ia – the religious leaders of Fire and Water. Historical books, such as *Phủ biên tạp lục* by Lê Quí Đôn (1726–1784), *Lịch triều hiến chương loại chí* (Calendar of Constitutional Type) by Phan Huy Chú (1782–1840) and *Đại Nam thực lục tiền biên* và *chính biên* (Dai Nam Records of the frontier and the main border) by the National Historiographer's Office of the Nguyễn Dynasty (1802–1945) consider these religious leaders to be leaders of kingdoms as well: the Kingdom of Thủy Xá (Water) and the Kingdom of Hỏa Xá (Fire) (Dournes, 2013).

For Pơtao Puih, the ceremonial prayer for rain is the most important ritual. In the area influenced by Pơtao Puih, the Jorai people believe that the sacred sword (called *tha* by the villagers) acts as a spiritual medium object and Pơtao Puih is the only being who can contact the gods to ask them for rain to water the upland fields. The ceremony to pray for rain is held in April at the side of the house of Pơtao Puih where the sun rises. At the end of the ceremony, people splash water around Figure 12.7.

The Ede people organize the river watering place ceremony annually after the harvest to express their gratitude to the Water Genie. The ritual is often held by the founder of the river watering place and presided over by the village old folks. All villagers enthusiastically and voluntarily take part in the ceremony, expressing gratitude to the Water Genie for giving them pure and abundant water sources. For the Churu people who live on wet rice cultivation, each village designates a person in charge of irrigation and two assistants. The head of the irrigation work is elected by the villagers. He must be good at irrigation and impartial. His task is to distribute equally the water supply from public canals to

Figure 12.7 Po'tao Puih the 15th in Gia Lai province worships the water god. (Photo by Nguyen Huu Dung).

each field of the village households. When necessary, he can ask the village head to mobilize villagers to renovate the irrigation works before the cultivation season.

So, the indigenous ethnic minority groups in the Central Highlands are always grateful to and respect the Water Genie. The role of the community in worship of the Water Genie is very important. They all consider it a sacred, respectful and gratitude-filled act. Worthy of note is the role of the shaman and the village elders in the ceremony to worship the Water Genie. Historically, in the Jorai and the Ede ethnic minority communes only, the shamans played the role of religious leaders of the Fire and Water Genies; a long time ago they were thought of as leaders of the Kingdom of Water and the Kingdom of Fire as recorded by Vietnamese historians. Friedrich Engels remarked: 'From the commencement of each of those communes, there were certain common interests whose conservation was assigned to every individual, but controlled by the collective such as judging disputes; punishing those who abuse power; and taking care of water sources…' (Engels, 1971).

12.4 THE WATER CULTURE OF VIETNAM THROUGH TWO ETHNIC GROUPS IN TWO CULTURAL REGIONS

The Kinh (the majority Vietnamese) in the Northern Delta and the indigenous ethnic minority group in the Central Highlands are two different ethnic communities living in two different spatial regions. Their behaviors toward water are different. The Kinh in the Northern Delta both fear water and desire to conquer it. Yet, the indigenous ethnic minority groups in the Central Highlands are respectful and grateful to water, considering it sacred. The water culture of Vietnam should be examined through historical evolution and cultural space with changes. In Vietnam, the ethnic group playing the role of the subject is the Kinh: 'Through ups and downs since the commencement of history, the ethnic group that will be the subject always settles in the delta while other ethnic groups settle in valleys or on mountain sides and the highlands. In the delta, people quickly turned to water rice cultivation with higher yield than subsidiary and upland rice crops in the uplands (Nguyen, 1996, p. 527).

The Kinh in the Northern Delta both fear and desire to conquer water related to their practice of wet rice cultivation. The legend of Son Tinh and Thuy Tinh (Mountain Genie and Water Genie) expresses the Kinh's idea of conquering water. The more the water rises, the higher the dykes will be embanked by them to prevent flooding. In 2008, at an international workshop on eco-culture of the residents living along the Red River from upstream to lower reaches, I presented a paper on the role and geographical and cultural position of the Red River dykes from the three-crossroads of the river to its lower reaches (Nguyễn, 2018, p. 912–934). The Kinh always consider water an environment of rituals in their village festivals. In their ceremony, people take water from the mid-stream flow of a large river to clean Joss statues. The statue in the sacred space of the monument must be cleaned with the water carried from the river to the center of the sacred space after a ceremony worshiping the River Genie and the Water Genie.

Of the 11 indigenous ethnic minority groups in the Central Highlands, only the Churu have behaviors toward water similar to the Kinh in the Northern Delta, although these two ethnic groups live in two different geographical and cultural spaces. They build canals, dykes and dams to provide irrigation for wet rice cultivation. The remaining indigenous ethnic minority groups in the Central Highlands behave with water differently to the Kinh in the Northern Delta. 'Here the village is not as attached to water as that in the Vietnamese society where the village is called làng nước' meaning village and water… In many regions, there are chieftains … at least there is a regime of religious leaders with the Fire King and the Water King' (Nguyen, 1996, p. 527). The annual ceremony to worship the Deities of the River Watering Place and the Water Genie after the harvest is a regular religious and cultural activity of the people of indigenous ethnic minority groups in the Central Highlands, whether they are Jorai, Ede or M'nong. The attitude of conquering water has not yet been seen in these ethnic groups.

12.5 CONCLUSION

The Kinh (the Vietnamese) in the Northern Delta and people of the 11 indigenous ethnic minority groups in the Central Highlands are two cultural subjects in two different cultural spaces in Vietnam. In the mind of the Vietnamese in the

Northern Delta and the indigenous ethnic minorities in the Central Highlands, there remains an attitude of respect and gratitude to water, a fear of water alongside striving to conquer it. The behaviour towards water of the ethnic groups in these two different cultural spaces in Vietnam have made the water culture of Vietnam unified but diverse, with different evolutions and changes in space and time which need to be further studied by Vietnamese and foreign anthropologists and cultural researchers.

REFERENCES

Dournes J. (2013). Potao, a theory of the power of the Jorai in Indochina, translated by Nguyên Ngọc. Knowledge Publishing House, Hanoi.

Engels F. (1971) Anti-Duhring. Su That (Truth) Publishing House, Hanoi, p. 304.

Gourou P. (1936). The Peasants of the Tonkin Delta. In: Vietnam Association of History Science, École française d'Extrême-Orient, Nguyễn K. Đ., Đào H. and Nguyễn H. O. (trans, 2003), Thế T. Đào (eds.), Trẻ Publishing House, Paris.

Nguyễn T. C. (1996). Contributing to Cultural and Ethnic Studies. Culture and Information Publishing House, Culture and Art Magazine, Hanoi.

Nguyễn T. (ed.). (2000). Village Conventions in Thái Bình Province. National Culture Publishing House, Hanoi.

Nguyễn C. B. (2018). Vietnamese Culture: Studies and Approaches. Social Sciences Publishing House, Hà Nội, Vol. **1**.

Trần T. (1984) Organizational Structure of Traditional Vietnamese Villages in Northern Vietnam. Social Sciences Publishing House, Hà Nội, p. 99.

Trần Q. V. (ed.). (the 20th edition, 2018). Basis of Vietnamese Culture. Vietnam Education Publishing House, Hanoi.

Section 5

The Current Practice of Water Cultural Education

Chapter 13

Unveiling Venice's waterways heritage. From the digital and extended Water Museum of Venice to UNESCO's Global Network of Water Museums

Eriberto Eulisse

ABSTRACT

The Venice is often associated with an image of rich commercial ventures eastwards and trade in the Mediterranean. However, it is worthy of note that for centuries Venice made huge investments westwards to modify the local hydrography and build a sophisticated network of inland waterways. The origin of this system can be traced back to the Middle Ages, when extensive hydraulic works were carried out to modify local watercourses and create artificial canals for commercial and strategic purposes. From the 16th century, improvements in agricultural production and a blooming river trade became possible thanks to functional hydraulic artefacts aimed at managing and controlling all water levels. This system of navigable waterways was carefully managed until the fall of Venice (1797), but subsequently, when railways and road transport came to dominate over inland navigation, it was abandoned. Today Venice's waterways and related waterscapes offer a unique setting to attract tourists. However, with the single exception of the Riviera del Brenta (that is renowned worldwide for its Palladian architecture), this heritage often lies in a worrying state of abandonment. After introducing the geo-historical context where careful management aimed at controlling all water flows was developed over the centuries, this chapter focuses on recent cultural dynamics of managing this inherited water asset. Considering the present state of neglected waterways there are also interesting opportunities for their revitalization. The use of digital tools combined with new eco-tourism strategies indeed can stimulate new models of water governance also for the

Editor: Xiao Yun Zheng
doi: 10.2166/9781789062045_0213

advantage of local communities. Today ground-breaking visions and new management approaches are needed to transmit to future generations this water heritage. In discussing the research strategy aimed at creating the 'digital and extended' Water Museum of Venice, this chapter will show how this model was useful to build a worldwide network of water museums.

Keywords: water heritage, water museums, waterways, waterscapes, sustainable development

> *Water, as with all things considered only for their usage, has become alienated from its own history, and made 'invisible'.*
>
> R. Franzin (2005), founder of Civiltà dell'Acqua

If the city of Venice is renowned for its centuries-old dominant position in Mediterranean maritime trade and its business and activities eastwards (Lane, 1973; Norwich, 1981; Zorzi, 2011), its huge investments and efforts aimed at controlling and modifying rivers westwards, and the impressive hydraulic works made to develop an efficient waterway system for inland navigation, are still largely overlooked. With the exception of a few rural and cultural landscape studies, the sophisticated inland waterway system of the Venice Republic which was built in the hinterland is definitely poorly studied in scientific literature (Ciriacono, 2006; Cosgrove, 1993; Vallerani, 2017).[1]

Inland waterways are definitely a 'minor' or secondary heritage, if compared to the architectural and artistic one developed in the island of Venice. Nonetheless, the substantial hydraulic heritage created by the city of Venice – or the *Serenissima*[2] – in its inland country estate is an interesting case study to reflect upon a neglected hydroscape and the considerable changes in local water management over the last few centuries.

Waterways have served for a long time as structural assets to develop intense trade-offs and as linear paths for socioeconomic growth and intercultural exchange. Thanks to fertile interactions among natural elements and human artefacts, even today it is possible to identify specific and unique water landscapes, or waterscapes along waterways (Schama, 1995).

It has been stressed that the emergence of Europe as a world power cannot be understood without recognition of the critical role played by its dense array of rivers and artificial navigable canals in industry, commercial trade and nation building (Hassan, 2010). Characterized by their intrinsic complexity, the European waterways go back at least to the Middle Ages, developed further during the Renaissance and reached their culmination during the industrial era, when rivers were transformed radically and new canals built to optimize water

[1]The reference here, in particular, is to the meagre scientific literature available in English.

[2]Deeds of the Venice Republic are often referred to in historical chronicles with the epithet of the *Serenissima* (i.e., the "most serene" or "unperturbed" city).

transport networks. In this process, the waterways even became a symbol of the modern nation-state building (Cosgrove & Petts, 1990; Swyngedouw, 2015; Vallerani, 2016).

Scrutiny reveals that the whole of Europe is characterized by dense stories related to interconnections among navigable canals and rivers which, together, make a unique waterscape heritage, or hydro-scape, with an intrinsic natural and cultural value, both tangible and intangible. It is only in recent years that surveying and cataloguing of waterways' heritage has become relevant to counter the silent but continuous loss of important aspects of the European cultural landscape. This is certainly the case for the minor or secondary hydraulic heritage which too often has been sacrificed on the altar of economic development 'at all costs' – a development aimed at setting up a progress that eventually proved to be not sustainable in the middle and long term.

In this chapter, following a research perspective outlined already (Eulisse & Visentin, 2018; Vallerani, 2017; Vallerani & Visentin, 2018), I will focus first on the geo-historical context of Venice's inland waterways heritage and then explore how new applications of ICTs (information and communication technologies) can help to recover an 'invisible' hydro-scape and generate an impact both at regional and international level.[3]

The aim of this investigation perspective is to understand the advantages to fostering sustainable tourism linked to particular hydro-scapes in regions that currently are not reached by tourist flows (Buhalis, 2003; Pease, 2007; Tang & Jang, 2010; Visentin, 2014; Willems & Schaink, 2015).

In the past decade, there has been growing focus in tourism studies related to rivers and historic waterways and canals. The multifaceted value of waterways – historical, socio-cultural and environmental – has stimulated increasing attention in the study of waterways' heritage and canal tourism from a management and marketing perspective (Chang, 2015; Prideaux & Cooper, 2009; Shipley, 1999; Vallerani, 2017; Vallerani & Visentin, 2018).

In this context, considering the existing state of decay and disrepair related to the minor heritage of Venice's waterways, from 2012 a project strategy was implemented by Ca' Foscari University of Venice and *Civiltà dell'Acqua* International Centre to sensitize both regional authorities and local communities about the potential tourism development of such a unique hydro-scape.[4] A few

[3]By the use of ICTs, I refer here not only to the process of digitization in the cultural heritage domain, but also to possibilities offered by the use of digital tools to enhance local awareness and strengthen ecotourism activities in order to promote a neglected water heritage.

[4]The *Civiltà dell'Acqua* ("Water Civilization") International Centre is a non-profit organization founded in 1996 and based in Venice, Italy. The Centre is active in promoting trans-disciplinary and holistic approaches to change unsustainable behaviours and practices in water perception, use and management. Its statutory goal is to settle a new, positive relationship between water and humans, that is to build a "new water culture". See www.civiltacqua.org and www.watermuseumofvenice.com [accessed on 15 February 2019].

projects funded by the European Union helped to support this applied research. These projects were aimed at stimulating new trends in preservation and new marketing approaches through the creation of a set of digital tours to promote eco-tourism and craft tourism activities.

Thus, the desire to explore and disseminate the multiple values linked to Venice's waterways heritage through digital media led to the creation of a number of itineraries co-designed with and by locals to rediscover and explore the forgotten hydro-scapes of the *Serenissima*. In 2016, at the 29th World Canals Conference, *Civiltà dell'Acqua* received the 'Innovation Award' for the digital project *Venice's Minor Waterways*, confirmed the growing interest towards the minor and secondary hydraulic heritage at international level.

The pilot project of the digital museum *Water Museum of Venice* was then initiated and gave new impetus to a research perspective that served as a model for the creation of the *Global Network of Water Museums* – an initiative endorsed, in 2018, by a dedicated UNESCO Resolution. This is an additional confirmation of the imaginative potential of innovative research approaches aimed at bridging, in a holistic perspective, past and present water civilizations with their related heritage and management models. Such a research perspective is fundamental to inspiring new approaches towards water sustainability education and water awareness efforts in the frame of implementing the UN's targets of Sustainable Development Goals (SDGs).

13.1 A HUMAN-MADE WATERSCAPE: THE INLAND WATERWAYS OF VENICE

An introduction to the history of Venice's inland waterways, their construction and pivotal importance to the blooming of the *Serenissima* Republic is essential to understand the context in which the digital and open-air museum Water Museum of Venice was designed, with the purpose to reinstate this valuable hydro-scape. The comparison between past and present water management models is implied in this approach and highlights the need to combine the preservation and the promotion of minor hydraulic heritage to the achievements of SDGs.

Venice's historic endeavours to control and divert different watercourses deeply transformed the local environment and, eventually, large portions of the regional landscape. Similar transformations required specialized technical knowledge which was the result of a peculiar socio-economic and cultural context, where the care for natural water flows and related hydro-morphology was always seen as a field of utmost importance by the Senate of the *Serenissima* Republic of Venice. Yet, despite the high level of sophistication of water management models and hydraulic works which have been developed over at least 700 years – from the 12th century to 1797, the rich heritage of Venetian inland waterways is today largely overlooked and neglected (Ciriacono, 2006; Cosgrove, 1993; Vallerani, 2017).

As shown by archaeological investigations, the present-day hydrography of the Venetian hinterland is the result of a long history of human interventions which dates back at least to Roman times. Historic cartography shows how Venice's hinterland featured both watercourses descending from the Alps (or 'alpine' rivers) and a considerable number of minor 'resurgence' rivers originating from the line of springs located between the high and low Venetian plains (Cosgrove, 1990; Vallerani, 2017). Many of these smaller but abundant streams eventually join the main alpine rivers, such as the Adige, Bacchiglione, Brenta and Piave rivers, which in turn discharge into the Adriatic Sea (Figure 13.1).

The Venetians carried out extensive hydraulic works in this watery hinterland in order to control and manage all rivers flowing into the Venice lagoon. In particular from the Renaissance to the end of 18th century, the city continuously modified and diverted all local water flows, including the mighty Po river, which was silting up the Venice lagoon with increasing sediment loads. These growing sediments represented a major threat to the city's survival as an island surrounded and protected by ever-changing waters. These hydraulic works and efforts created a unique heritage and an unmatched chapter in hydraulic history: a human-made waterscape, which is peculiar to the entire hinterland of Venice (Ciriacono, 2006; Vallerani, 2017).

Due to irregular flow patterns and differences in water levels, not all local rivers were naturally suitable for navigation. Fluvial sections which permitted safe river navigation were quite limited. For this reason, from the 12th century onwards a number of artificial canals were constructed to strengthen fluvial trade connections among Venice, Padua, Bologna, Ferrara, Mantua and Milan, to name a few. As a result, an extensive network of navigable waterways was created to facilitate both trade and the strategic transportation of military arms (Vallerani, 2017).

Since Medieval times, local rivers were seen as part of a complex hydrography where natural watercourses could coexist with a number of artificial canals. However, since waterway construction implied considerable investments and complex hydraulic works, only some independent city-states in the Venetian region, such as Padua, Vicenza and Treviso, which are located not far from Venice, could afford it.

The Venetian waterway system was developed initially in the first half of the 12th century, when the two rival fluvial cities of Padua and Vicenza underwent a period of significant expansion. Both these city-states were built along the banks of the Bacchiglione river, as they considered control over the local waterways a strategic asset to defend their urban communities and supply them via the efficient system of a fluvial-based commercial trade (Eulisse, 2014; Vallerani, 2017).

In this period the inhabitants of Vicenza, due to their location upstream of Padua, decided to divert most of the runoff of the Bacchiglione river into an artificial canal, in order to bypass the rival city downstream. According to historic chronicles of the time, the new canal's excavation was completed in 1139, thereby making navigation possible from Vicenza directly to the southern lagoon of Venice and to the

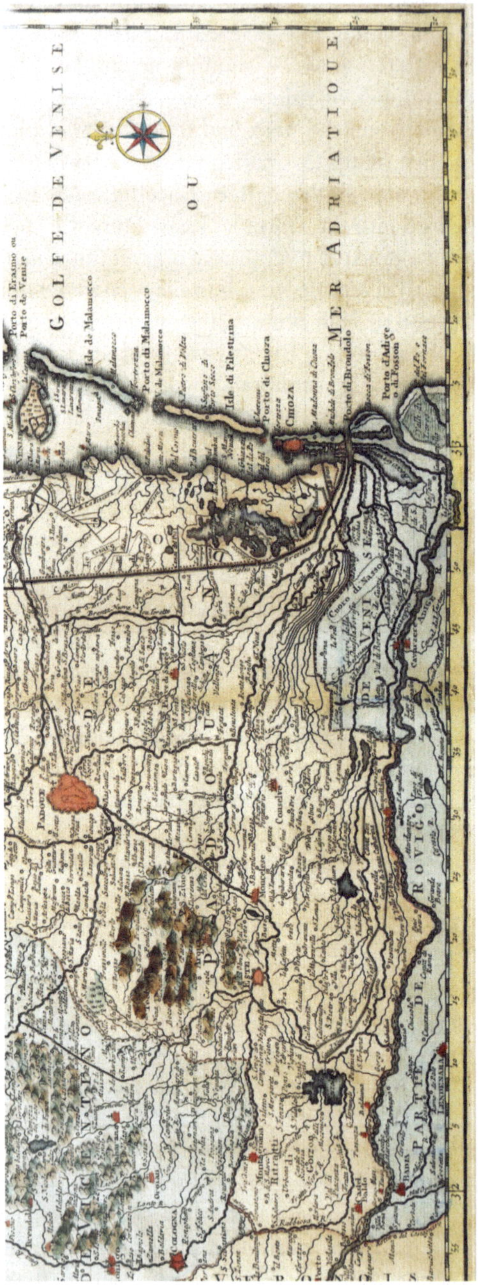

Figure 13.1 Eighteenth-century map of the Duchy of Venice (the Dogado) by Paolo Bartolommeo Clarici and Paolo Cantelli where the inland waterways system of Venice (the fish-shaped island on the top) made by the main alpine rivers, the smaller or 'resurgence' rivers and the artificial canals can be easily identified. Image Courtesy of the municipal library of Battaglia Terme (Padua).

productive saltpans of Chioggia without crossing Padua. A few years later, also Padua completed the construction of two important canals to speed up its 'liquid connections' to Venice: the Battaglia Canal (1201) and the Piovego Canal (1209).

After the fall of Constantinople, in 1476, the interests of the Venetian aristocracy increasingly shifted towards the mainland located westward and its waterway system. Huge capitals were invested in order to develop a 'modern' agriculture along waterways. This period was characterized by a number of massive hydraulic works, which included drainage management, the further development of a complex artificial system of navigable canals and the diversion of major and minor rivers away from the Venice lagoon. The need for an overarching governance of the waterway system of the entire region was felt strongly from this period by the Venetian aristocracy, since it become clear that continuous transport of river sediment loads was encouraging the transformation of the Venice lagoon and its islands into unsafe marshes and, eventually, into a wretched mainland without natural defences (Ciriacono, 2006; Eulisse, 2014; Vallerani, 2017).

From the Renaissance, the Venetian Senate become aware that direct governance of its natural defence system – the amphibious environment of its lagoon – and a thorough hydraulic management plan was urgently needed in all its surrounding areas: from the Tagliamento river to the Po river, along a coastline of approx. 200 kilometres. This awareness led to many of Venice's 14th- and 15th-century military interventions. Because of this geopolitical strategy, by 1484 a vast territory, approximately corresponding to the present-day Veneto and Friuli regions, become the new inland possession of the mighty sea-faring city-state of Venice (Ciriacono, 2006; Lane, 1973; Norwich, 1981; Zorzi, 2011).

To better manage its lagoon environments and related watercourses, in 1501 Venice established the magistracy of the *Savi alle Acque* ('Water Masters'), which then became the *Collegio alle Acque* and finally, a few decades later, the *Magistrato alle Acque* ('Judicial Water Council') – an institution which still operates today in spite of its many transformations.

From the 16th and 17th centuries onwards, Venetian hydraulic engineers increasingly focused their efforts on preventing all freshwater from flowing into the lagoon in order to avoid silt deposits. A number of renowned *Savi alle Acque*, including Cristoforo Sabbadino (1489–1560), embarked between 1601 and 1684 on a series of public works to divert systematically all watercourses flowing into the lagoon, including the Brenta, the Sile, the Po and the Piave rivers. Fluvial diversions were made in conjunction with the creation of new reclaimed lands to improve agricultural production. In 1556, also the magistracy of the *Beni Inculti* ('Uncultured Lands') was established with the aim of making new land. Innovative techniques for drainage, irrigation and water management were adopted. Both magistracies of *Savi alle Acque* and *Beni Inculti* are proof of the city-state's policies and efforts to manage carefully the local hydrography. The extensive works carried out by these magistracies produced an impressive number of books and maps, which is still available in the central archives of the Venice Municipality (Ciriacono, 2006; D'Alpaos, 2010; Tiepolo & Rossi, 2008).

Beyond trade opportunities, waterways also became more and more popular as sites for leisure for the Venetian aristocracy, who from the Renaissance started to commission construction of farms, mansions and, eventually, magnificent villas and pleasure gardens – thus creating unique waterscapes around them. Side canal locations were listed among the most attractive sites in the mid-16th-century treatise of Andrea Palladio on architecture. A good example of this is known today as the *Riviera del Brenta*. This canalized former stretch of the Brenta natural river connects the Portello, the river port of Padua, to Venice and its lagoon, and is of strategic importance not only for river trade. The *Riviera del Brenta* waterway is also known as the 'elongated Grand Canal' of Venice, since the 40-kilometre artificial canal is lined with amazing buildings, such as villa Pisani, villa Malcontenta and villa Contarini in Piazzola sul Brenta, to name a few (Figure 13.2). Initially manors were built to support new farms for agrarian experimentation and to improve agricultural production, but many of them developed then into elegant villas for aristocracy entertainment, leisure and as symbols of power. As such, thanks to well-tested hydraulic management models and to secure waterways connections, they were used for three centuries until the fall of the Republic (Eulisse, 2014; Vallerani, 2017).

After the fall of the *Serenissima* (1797), the complex system of Venetian waterways remained in use until the First World War. Between the two world wars, it was still used in spite of the increasing competition with railways. In the 1950s and 1960s, however, waterways trade declined dramatically. Neglect and indifference towards this impressive hydraulic heritage prevailed. From the 20th century, new development policies aimed at strengthening only railway and motorway networks at national level became the norm, and reduced dramatically all investments previously directed to maintain the system of 'liquid roads'. By the end of the 1960s, inland navigation trade was definitely abandoned, and this decline was paralleled by the rapid functional disrepair of most of the hydraulic infrastructure (Eulisse, 2014; Vallerani, 2017).

This decay had large effects on cultural and social attitudes, as well as on economic and artisanal activities. It marked the loss of the memorial value of Venetian inland waterscapes. However, after half a century of abandonment and decay, today a new awareness as arisen from the recognition of both their cultural and natural value, including ecosystem assessment. This awareness is crucial to develop new attitudes and behaviours towards the substantial cultural heritage related to historic waterscapes, with the goal of rehabilitating hydraulic landmarks and the fascinating vestiges left by the *Serenissima* in a unique landscape.[5] In this

[5]As a result of this growing awareness, from the 1990s a considerable number of books focusing on history and features of Venetian rivers and waterways have been published, in particular by Cierre Publishing (http://edizioni.cierrenet.it/volumi/).

(a)
(b)

Figure 13.2 Two sites of the digital and open-air Water Museum of Venice: (a) the Palladian architecture of Villa Contarini (Piazzola sul Brenta), and (b) the river port of Portello (Padua) which was celebrated in Canaletto's paintings. Photo courtesy: Civiltà dell'Acqua image archive.

perspective, the hydro-scapes of Venetian waterways suggest the potential to develop new eco-tourism and craft tourism activities which, nonetheless, until today received only a limited consideration by water management policies at regional level.

13.2 NEGLECTED HERITAGE, SENSE OF PLACE, ECO-TOURISM AND THE ROLE OF LOCAL COMMUNITIES

Venice's long history of hydraulic engineering unveils the peculiar amphibious features of a large hinterland area and, as such, its potential to develop new functionalities and, in particular, new eco-tourism activities in which the active role of local communities is fundamental to redeem the neglected waterways heritage. In this perspective, the evolution of Venetian waterways and related hydro-scapes can be compared with those affecting the waterways of central and western sections of the Po Valley, towards Milan, as well as with other European regions. Indeed, both Lombardy and Emilia Romagna have a remarkable historical and cultural heritage related to inland waterways which, although they have ceased definitely their function and relevance in fluvial trade, have a great potential to support new eco-tourism activities and new recreational uses for residents.

The waterways network of Northern Italy from Milan to Venice reached its height in the early 20th century, when it was developed further during First World War to support the Italian troops on the eastern front against the Austro-Hungarian Empire. At the time, the entire navigable network included the regions of Veneto and Lombardy and spanned about 1,700 kilometres from Venice to Milan, with different connections to Locarno, Mantua, Ferrara, Padua and Trieste (Cucchini, 1931; Vallerani, 2016). This waterways network is today managed as a self-contained system with fragmented regional policies, regulations and development approaches.

After the Second World War and since the Italian Republic was established (1948), policies dedicated to waterways at national level ceased absolutely. Today, the regional governments of Veneto and Lombardy seem uninterested in developing an overall management plan to support the recovery of waterways' heritage and navigation for tourist purpose (with the exception of the Riviera del Brenta, the well-known elongated 'Grand Canal' of Venice and the *Navigli* of Milan), despite their considerable potential as tourist attractions.

Unlike other European countries such as England, France and the Netherlands, where in the 1980s and 1990s the number of leisure boaters increased significantly and favoured the transition of the waterways system from a transportation network towards leisure and tourism, in Italy this transformation has proven to be more difficult, mostly because of highly fragmented or non-existent policies. A large scientific literature is available on the above European countries.

However, the waterways navigation system created by the *Serenissima* suffers not only from functional decay but simultaneously also from the loss of a 'fluvial sense of place', that is, the loss of perceptions and attitudes that for centuries inspired so many people's life and behaviours and which were truncated in the 1960s, with the sudden stopping of traditional river navigation and trade. Such a loss is testified also in many books of local history which investigate the tangible and intangible culture linked to the tradition of Venice's inland navigation.[6] Also ancient trades and crafts related to traditional wooden boats are considered by historic and ethnographic literature to collect and recover the material culture, craftsmanship and habits related to past inland navigation.

Today, after half a century period of decay, a spontaneous revaluation of inland waterways' heritage seems to be occurring due to raised (although sporadic) awareness among riverine communities of their 'sense of place', which is traditionally based on long-standing life styles shaped by material relations with local rivers and water courses. Practical examples include the rowing clubs devoted to the traditional *voga alla veneta*, [7] and other popular events, such as the so-called *Vogalonga Euganea*, or the *Remada a Seconda*.[8] These are among the few existing attempts to keep alive the local rowing traditions and the 'amphibious identity' of many small river towns (Pergolis & Pizzarello, 1981).

Riverine communities and associations also show an increasing interest in the ecological conditions of riparian environments, as far as they provide ideal settings for sports and recreational opportunities for residents and help to improve the quality of daily life. In this context, water management issues should be addressed by new governance strategies aimed at safeguarding and restoring the natural and cultural environments, and simultaneously at satisfying the growing demands of residents and citizens for attractive waterscapes to relax in and regenerate physically and psychologically in their daily life.

Inevitably also the water management dynamics of less-known and minor waterscapes will need more attention from policy makers. In the Veneto region, some of them have become increasingly sensitive to identify the potential water features of their territories as an ideal asset for cultural tourism, as the river towns of Battaglia Terme, Casale sul Sile and Portogruaro show.

To avoid the complete loss of memories and habits linked to the fruitful relationship between humans and waterways, it must be stressed that local hydro-scapes deserve more attention also from academia and researchers, who can play an important role in regenerating the local sense of place linked to inherited hydro-scapes.

---

[6]Among relevant publications on the subject: Zanetti, P.G. (ed.) 1998 *I mestieri del fiume. Uomini e mezzi della navigazione [The river's crafts. Men and trades of navigation]*, Cierre Edizioni, Verona, Italy. Zanetti, P.G. (ed.) 1998 *La navigazione fluviale e il museo di Battaglia Terme [Inland navigation and the River Museum of Battaglia Terme]*, La Galiverna, Battaglia Terme, Italy. Pavan, C. 1989 *Sile, alla scoperta del fiume [Discovering the Sile river]*, Cooperativa Servizi Culturali, Treviso, Italy. Jori, F. 2012 *L'ultimo dei barcari. Riccardo Cappellozza, una vita sul fiume [The last boatman. Riccardo Cappellozza, a life on the river]*, Tipografia Sartor, Pordenone, Italy. Penzo, G. 2005 *Maestri d'ascia. Costruire barche a Venezia [Shipwright. Building boats in Venice]*, Marsilio, Venezia, Italy.

[7]The *voga alla veneta* is a typical rowing technique used with specific wooden river boats, where the rower paddles standing up from the bow.

[8]The *Remada a seconda* is a funny boat race, recalling the *Regata Storica* (historic regatta) of Venice, but in a river context.

It remains mainly to local communities and policy makers to take advantage of the new possible development strategies linked to a kind of tourism related to their traditional hydraulic heritage and water culture. It is definitely up to them to exploit all opportunities provided also by digital media to promote and enhance new approaches to eco-tourism and boost, with this strategy, the rural economy.

13.3 HOW DIGITAL MEDIA CAN HELP TO IMPROVE ACCESSIBILITY AND REDEEM THE INHERITED HYDRAULIC HERITAGE

In recent decades, the application of ICTs to cultural heritage and tourism activities has been driven by growing interest in the opportunity it provides to enhance meaningful experiences in the tourist sector as well as in research in the arts and humanities (Arnold & Geser, 2008; Haus, 2016).

According to scholars, applications for smartphones and tablets are revolutionising tourists' experience of the travel domain. The increasing number of applications that are being produced constantly is changing our relationship with travel and its organization, as well as our sense of orientation and mobility. The use of interactive maps, digital information and databases are transforming our habits concerning the ways we perceive space and place, as well as our relationships with people we meet when travelling (Dickinson, 2014; Wang, 2012).

ICTs transformed definitely the way we look at and interpret the world. In the field of tourism, for example, ICTs increasingly influence the way we travel and our approach our destination with cognitive and artificial intelligence applications, new semantic representations, space-oriented applications (GIS) and virtual reality applications. These digital applications affect both the quantitative and qualitative aspects of travel for all users in many and often unpredictable ways.

The fertile relationship between cultural heritage and digital technology is a productive field of investigation (Kalay, 2008; Neuhofer, 2012; Xiang & Tussyadiah, 2014). Also the creation of a number of journals dedicated to digital technologies and cultural heritage confirms this growing research interest.[9] Due to development of user-friendly software, an increasing number of scholars use ICTs for their research into heritage. Also the use of information technology for heritage planning has become a research objective in itself (Eulisse & Visentin, 2018).

Research into cultural geography and media studies has thrown light on different impacts of new technologies in refashioning our knowledge of places to visit. However, there are some basic distinctions to be made among different digital

[9]See for example *International Journal of Heritage in Digital Era; International Journal of Human-Computer Studies; New Media and Society.*

media and their effective potential in shaping 'meaningful experiences' for tourists. Indeed it has been stressed that while digital cartographies, such as Street View or Google Earth, tend to obliterate place by providing functional images where real life context is overlooked, in contrast geo-spatial technologies can produce a quite different experience, and provide residents and visitors with a deeper understanding of place and related context (Gilchrist, 2017). In this perspective, the use of geo-spatial technologies aimed at promoting direct and physical access to cultural heritage is a prerequisite to create meaningful experiences for visitors. This is quite different from digital uses where only 'virtual experiences', made just with digital media and without physical access to heritage, are considered.

In this sense, ICTs are a powerful tool to create and emphasize an 'inclusive' vision of heritage. As noted in the communication of the European Commission *Towards an integrated approach to cultural heritage for Europe* 'digitisation multiplies opportunities to access heritage and engage audiences. [...] The digitalisation of heritage contributes to improve public access to different forms of cultural and linguistic expression. Digitising cultural heritage, making it accessible online, and supporting its economic exploitations are activities at the heart of the Digital Agenda for Europe' (EC, 2014). The importance of the digitization process in improving access to different forms of cultural heritage and the role that local communities can play in this process to favour and support the development of a sustainable tourism are the core aspects of this communication.

The interplay between hydraulic heritage and geospatial technologies (geo-ICT), aimed at improving accessibility, stimulating civic engagement and influencing policy agendas to develop a 'slow' tourism, forms an interesting case study of Venice's waterways (Eulisse & Visentin, 2018) and, for the purpose of this chapter, enables us to highlight different cultural dynamics of local water management through the last centuries.

Stereotyped perceptions towards neglected waterscapes and water memories related to minor heritage are a stimulating field where to test how digital media can challenge the visitor experience in their travel domain, by offering alternative routes for a growing number of new interests. Mobile technologies, for example, can be used to improve access towards different forms of secondary hydraulic heritage and offer alternative and meaningful experiences. Different types of sustainable mobility can be considered in order to satisfy the growing tourist expectations for new kinds of experiences. For example, new digital routes can be made by kayak, boat, bicycle, horse and on foot, and the same itineraries can also serve rural tourism networks, hikers and cycling associations.

In this perspective, strategic research applied to Venice's waterways provided both the opportunity to start digitizing the minor hydraulic heritage, and to improve public access towards derelict and neglected waterscapes. The final output was the creation of a digital and open-air museum, aimed at linking the most relevant monumental heritage and places of the water civilization inherited from the *Serenissima* Republic of Venice.

13.4 FROM THE DIGITAL AND EXTENDED WATER MUSEUM OF VENICE TO THE GLOBAL NETWORK OF WATER MUSEUMS ENDORSED BY UNESCO-IHP

Looking at the small city centre of Venice which nowadays is visited by some 30 million tourists annually, the challenge of developing a sustainability frame of reference to reduce the considerable impacts of mass tourism goes side by side with the creation of 'alternative' itineraries aimed at offering new and different experiences in the field of tourism. In this frame, also Venice's minor hydraulic heritage can play an important role to convey messages related to sustainable mobility and SDGs.[10] Alternative routes can include visits to hydraulic artefacts along derelict waterways and to vestiges of traditional river navigation with related craftsmanship. In this perspective, different isolated areas of the Venice lagoon can be reconnected again to the hinterland.

The Venice Lagoon includes nine municipalities and more than one hundred small and 'invisible' islands. Today, in spite of mass tourism – or maybe because of it – the lagoon is suffering from heavy, rapid and uncontrollable depopulation. 'Experiential' tourism could be therefore a strategy to open forgotten and hidden islands such as Burano, Pellestrina, Sant'Erasmo and Lazzaretto Nuovo, to name a few, to new experiences and offers related to craft tourism, fish tourism and genuine local produce.

In this context, Venice's hidden waterways may serve as a linear blue frame for alternative itineraries and to launch new destinations for more sustainable tourism, connecting the city centre to neglected areas of the lagoon and to other peri-urban inland contexts, where abandoned hydro-scapes (including natural and cultural, tangible and intangible heritage) which are not contaminated yet by mass tourism can let tourists discover and experience an unconventional, 'forgotten Venice'.

The second part of this chapter focuses on how a project strategy was built to redeem the derelict heritage of Venice's historic waterways through the use of digital media and the involvement of local communities. From 2012 to 2018 such a strategy was launched and developed by a solid partnership among Civiltà dell'Acqua International Centre, Ca' Foscari Venice University and the UNESCO Venice Office.[11] As a concrete result of this cooperation, two main outputs were produced: the digital and open-air Water Museum of Venice and the Global Network of Water Museums which, in 2018, was officially endorsed as a 'UNESCO initiative' (Civiltà dell'Acqua International Centre, 2014; Global Network of Water Museums, 2018).

In 2012 a seminal conference titled 'Memory, Identity and Responsible Tourism along European Waterways' was organized in the river town of Battaglia Terme, close to the Euganean Hills and Padua. Once this town was a flourishing crossroad

---

[10]See: https://sustainabledevelopment.un.org/?menu=1300 [Accessed 5 January 2019].

[11]Regional Bureau for Science and Culture in Europe.

of inland waterways visited and portrayed by notorious travellers, such as the French philosopher Montaigne. Today, in contrast, it is just a minor and invisible river town. A town, however, which deserves to be explored for its hidden and amazing hydro-scape vestiges, including the renaissance Castle of Catajo and the baroque monumental garden of Villa Barbarigo in Valsanzibio (Figure 13.3).[12] The conference was a unique opportunity to meet different European partners interested to protect and promote, in particular, the heritage of minor waterways. It generated the visionary Manifesto titled 'European Charter of Minor Waterways', to bring more institutional attention towards this neglected but relevant hydro-scape.

In 2013 Civiltà dell'Acqua started to develop, in cooperation with a team of geographers of Venice University and thanks to some funds granted by the European Union, a research project aimed at digitizing the heritage of Venice's derelict waterways and create new digital routes to promote sustainable eco-tourism in less developed rural areas. The project goal was to foster local economies by stimulating visits to this neglected hydraulic heritage through mobile applications. Between 2013 and 2015, four mobile applications and 60 digital routes dedicated to Venice's minor waterways were named just as many 'liquid roads' have been used historically by the *Serenissima* Republic for river navigation. All applications are downloadable for free, in both Italian and English, and include: 'Tagliamento River' (with the navigable river segment linked to the major waterway that once connected the ancient Roman towns of Ravenna and Aquileia); 'Delta Po River' (aimed at exploring the wild and amazing branches of the Po river with their impressive naturalistic setting); and finally 'Bacchiglione River' and 'Waterways of the Euganean Hills' – both referring to relevant segments of Medieval 'liquid roads' which connect the hinterland of Padua to the Venice Lagoon. All applications were also used as pilot tools to involve local communities and associations to rediscover their water heritage and environments, and regenerate the local 'sense of place'.

In 2015, the project of the digital and open-air Water Museum of Venice took shape in the frame of EXPO Venice,[13] thanks to generous private contributions, and as a combination of most relevant past and present water heritage, both cultural and natural, also including present-day best practices of integrated water management. An example of this peculiar mix, which today includes more than 70 relevant places of water civilization, are the Botanical Garden of Padua's University, listed in the World Heritage List of UNESCO, and the lagoon paths of fishing basins in the Po Delta (Figure 13.4).[14]

[12]*Catajo* in the Venetian language literally means the "mansion of the excavated canal", that is, the building which was functional to control river trade and gradually evolved into a magnificent castle during the Renaissance period.

[13]Expo 2015 is the title of the universal exposition hosted by Milan from May to October 2015. Venice was officially the only Italian city officially twinned with Milan through the programme "EXPO Venice".

[14]Foundation CARIPARO (Cassa di Risparmio di Padova e Rovigo).

(a)

(b)

Figure 13.3 Two sites included in the digital and open-air Water Museum of Venice: (a) the Catajo Castle built to control the Medieval waterway system of the Euganean hills, and (b) the monumental garden of Villa Barbarigo in Valsanzibio. Photo courtesy: Civiltà dell'Acqua image archive.

The digital and open-air Water Museum of Venice (www.watermuseumofvenice.com) was launched in conjunction with EXPO with a combination of different 'immersive installations' and itineraries of Augmented

(a)

(b)

Figure 13.4 The digital and open-air Water Museum of Venice: (a) the Botanical Garden of Padua listed in the World Heritage List of UNESCO as 'the origin of all the world's Botanical Gardens', and (b) the digital route along the lagoon fishing basins in the Po delta (Rosolina).

Reality along Venice's invisible waterways. The multi-sensorial space 'Exploring the Venice Lagoon' was hosted for eight months in Palazzo Zorzi, at the premises of the UNESCO Venice Office, to explore the different forms of water civilization to be found in the entire Venice Lagoon. The interactive installation 'Memories of the Po Delta', which today is still operational, was aimed at

Figure 13.5 The interactive installation of the video artist Paolo Scoppola, *Aquaboll*, created for the exhibition 'Visualizing Water Worlds', at EXPO Venice 2015, was functional to launch the Water Museum of Venice.

reconstructing the water worlds of river navigation along the Po River. The interactive installation 'Visualizing Water Worlds' was aimed at investigating different issues related to sustainable water uses and took place at the main entrance of EXPO Venice, in the 'Tesa 105', in combination with stimulating video games to attract children and visitors to discover the 'liquid shapes' of their bodies (Figure 13.5).[15] In addition, two itineraries of augmented reality aimed at exploring the substantial heritage of nautical history related to local waterways were made in the river towns of Jesolo and Battaglia Terme.

In this frame, the digital platform of the Water Museum of Venice was conceived both to promote the digital installations made for EXPO in different locations and as a repository of digital heritage representations of the whole area crossed by Venice's historic waterways: from the Tagliamento river to the Po river, along a coastline of approx. 250 kilometres, including on the western side the entire Venetian hinterland up to the Italian Alps.

In 2016, international recognition of this project strategy focusing on derelict waterways and hidden hydro-scapes was the awarding of Civiltà dell'Acqua with the *Innovation Award* at the 29th World Canals Conference, in Scotland (Figure 13.6).

The methodology developed to implement this early project phase served as a foundation to implement the project 'European Waterways Heritage' in 2017, with funding from the JPI-CH programme of the European Union to create the

[15] *AquaBoll* is the title of the digital installation made by the video artist Paolo Scoppola to capture the fleeting attention of EXPO visitors and children with "liquid" interactive games.

Figure 13.6 The awards ceremony at the 29th World Canals Conference, in Scotland (2016), when *Civiltà dell'Acqua* was granted with the 'Innovation Award' for the digital project on Venice's Minor Waterways.

digital platform 'Waterways Explorer'.[16] The project, leaded by Ca' Foscari Venice University, was aimed at generating an additional series of digital itineraries focusing on a new set of secondary waterways in a comparative and trans-boundary perspective. A refined methodology was co-developed by four European countries in order to create a number of geo-mapped itineraries supporting video interviews and other digital media in an open source platform.[17]

The advanced methodology that was generated improves understanding of how digital media were conceived as a tool to involve local communities, stimulate eco-tourism and re-create a new 'sense of place' along abandoned cultural waterscapes.

The first research activity included a digital census of the relevant tangible and intangible heritage linked to hydraulic heritage along the 'liquid roads' of each pilot area in the United Kingdom, Spain, Italy and the Netherlands: from Liverpool's and Manchester's canals which were used during the industrial revolution to the irrigated landscape of Girona and its canals close to the

---

[16]See: www.waterwaysexplorer.org [Accessed 5 January 2019]. Project leader: Ca' Foscari Venice University (Italy); project partners: University of Brighton (UK), Leiden University (Netherlands) and University of Girona (Spain). Associated partners: Canal and River Trust (UK), Waterrecreatie Nederland (Netherlands), Consorzio di Bonifica 'Acque Risorgive' (Italy), Campus Euromediterrani del Turisme i l'Aigua (Spain), and Centro Civiltà dell'Acqua (Italy).

[17]See the "storytelling platform" for mobile phones Izi.Travel.

Pyrenees; from Venice's waterways to the Dutch fortified town of Enkhuizen, the wind mills of Nieuwkoop and the 1000 islands of Broek op Langedijk.

Collection of data proceeded from a review of the available literature (including books, local newspapers and journals) and through interviews with local historians and associations. All planned routes were tested and mapped with Google Earth in KMZ files and, when possible, co-designed with the local communities. During the fieldwork, associations were sensitized and involved for final shaping of the storytelling, again through video interviews. All oral stories were then incorporated in the same digital routes to reflect riverine communities' interests and inputs (Eulisse & Visentin, 2018).

From the very beginning, both locals and tourists were considered as the targeted end-users of research investigation. Among the main objectives there was not only the need of increasing the value of minor rural areas crossed by waterways for the inhabitants, in order to revitalize a weakened or lost sense of place, but also the necessity of supplying tourists with a number of free digital devices to improve access towards the invisible hydro-scapes. In this perspective, the project was developed in a twofold direction: on the one hand, it paid attention to the interests of outsiders – the tourists – and on the other it focused on insiders, considering local memories and the related socio-cultural texture inherent to both tangible and intangible heritage.

Indeed, a thorough investigation of the local water heritage cannot be limited to material hydraulic structures, but must include values, perceptions and behaviours associated with it. In order to understand the cultural diversity linked to specific waterscapes, it is therefore necessary to consider also the intangible heritage with its traditional knowledge, oral traditions, vernacular languages, craftsmanship, arts, culinary traditions and specific forms of social organization (Smith & Akagawa, 2008; Vallerani & Visentin, 2018; UNCCD 2005).

The project *Waterways Explorer* demonstrated that the heritage of minor waterways has a similar cultural potential as the heritage of major ones, although it is more distant from the national and European levels. At local level, cultural identity tends to be more localized and is thus invisible to major policy agendas. On the other hand, it is also less affected by standard approaches. For this reason, the tourism potential tends to be more focused on small-scale ecotourism and leisure activities, rather than on mass tourism. In this sense, secondary and minor waterways appear to be essentially local waterways for local people, although they do have the potential to fulfil many functions of the major waterways, including attraction of new and consistent tourist flows (Eulisse & Visentin, 2018).

In 2018, the strategy to redeem Venice's invisible hydro-scapes was further implemented with two additional projects. In the frame of the Interreg project 'YouInHerit', five new digital itineraries on craft tourism related to nautical heritage were developed by Civiltà dell'Acqua in cooperation with the Venice University. The aim of the project was to involve young generations towards traditional craft heritage.[18] Students were trained for a period of eight months to co-design digital tours and audio guides with locals in order to explore the 'forgotten Venice', that is, the invisible nautical heritage and the related craftsmanship in five different *sestieri* (districts) and islands of Venice (Figure 13.7). All digital tours can be explored by use of interactive maps that group together different kinds of knowledge (perceptive, spatial, cultural and geographic). Training of young people to co-design new digital trails on nautical heritage proved to be successful in order to sensitize also young tourists towards the forgotten heritage of traditional wooden boats.

In total, from 2012 to 2018, some 90 digital itineraries were developed by *Civiltà dell'Acqua* with the digitization of almost 500 locations, or POIs ('point of interest'), related to minor hydro-scapes. The proposed tours to different locations range from visits to castles, water-mills, river ports, locks, wooden boats, rural farms and elegant villas. Some of them also include experiences with genuine local produce or visits to local craftsmen.

A final, major output of the project strategy which started in 2012 is the newly created Global Network of Water Museums, which groups together a growing number of different water museums worldwide. In May 2017, the first International Conference of the Global Network of Water Museums was organized in Venice with the support of UNESCO Venice Office. Its purpose was to test the possible creation of a 'digital and open-air museum', similar to the model of the Water Museum of Venice described above, but at global level. Some 30 museums and institutions from all over the world participated in the conference and confirmed interest in creating an institutional network. In June 2017 a proposal was made to the 55th Bureau meeting of UNESCO-IHP (International Hydrological Programme) in Paris to establish the network under UNESCO's auspices.

In June 2018, the UNESCO-IHP Resolution n. XXIII-5, titled 'Global Network of Water Museums and UNESCO-IHP in Support of Water Sustainability Education and Water Awareness Efforts', was officially endorsed by the Intergovernmental Council of the International Hydrological Programme of UNESCO. Recalling previous UNESCO's recommendations, this Resolution endorses the Global Network (WAMU-NET) 'as an IHP initiative to create synergies within UNESCO with the aim of better coordinating water museums worldwide and of improving water management via communication and education activities'. This was definitely a crucial output which blessed the innovative research strategy described in this paper at the international level (Figure 13.8).

The opportunity to share the methodological approach elaborated by the Venice University and *Civiltà dell'Acqua* with WAMU-NET's members was further achieved through the Great Rivers Forum: the international conference which in October 2018 was organized in Wuhan, China, by UNESCO Beijing Office at the

[18]See https://www.interreg-central.eu/Content.Node/YouInHerit.html [accessed 15 February 2019].

Figure 13.7 The digital itineraries 'Exploring forgotten Venice' (Interreg project YouInHerit) aimed to involve young people in co-designing a number of routes on traditional sailing and craft tourism: (a) video interview with Gigi Divari on traditional fishing, and (b) with Saverio Pastor, one of the last Venetian craftsmen producing traditional oarlocks for gondolas.

premises of the Changjiang (Blue River) Civilization Museum. A conference session titled 'Visualizing Water Worlds: Heritage, Museums, and Sense of Place' offered the opportunity to reflect on how water museums can support new representations of traditional river cultures in their displays and simultaneously a 'new water culture' linked to the implementation of SDGs' targets. Session

Figure 13.8 The team of the Global Network of Water Museums after the approval of UNESCO-IHP Resolution n.XXIII-5 on 15 June 2018 in Paris.

participants highlighted from different perspectives the potential of cooperation among water museums, education centres and different UNESCO programs and networks to strengthen the current debate on sustainable development goals to support water awareness efforts.

Indeed water museums can play a crucial role in creating new bridges between science and society. If their core activities are focused on public engagement, education and dissemination, on the other hand they can also support and contribute to research activities using their archives and collections to provide deeper understanding of the local water heritage (Figure 13.9). In this sense, the session highlighted how best practices, related to integrated river management and sustainable water uses, linking past and present good practices, need to find new ways of visualization in museum displays. Also attractive visualizations of traditional water worlds in a comparative perspective can provide comprehensive anthropological understanding of the different cultural attitudes and traditional knowledge related to rivers and water uses.

Museum displays need constant renovation to be appealing to the public and especially to young generations. Today, the use of digital media and interactive tools is fundamental to engage better the audience and influence people's perceptions, attitudes and behaviours about the intrinsic 'value' of water and its deep association to SDGs. Research activities should be aimed at filling the present gaps of stereotyped approaches towards the richness of meanings culturally associated to rivers and waters, as well as the inherent diversity of water cultures all over the globe. In this frame, also new curricula and capacity

(a)
(b)

Figure 13.9 Two museum members of UNESCO's Global Network of Water Museums: (a) water games for children at the Yaku Parque Museo de l'Agua in Quito (Ecuador), and (b) the reconstruction of fluvial ecosystems along the Blue River at the Changjiang Civilization Museum of Wuhan, China.

building programmes for museum staff, educators and trainers should be developed to improve education activities in line with SDGs.

To implement SDGs in a holistic perspective, there is the urgent need to involve youth and provide them with a role to play in establishing a 'new water culture' linked to both their inherited water heritage and to SDG n.6 ('clean water and sanitation'). Indeed water is not just a chemical formula.

In this perspective, water museums should co-operate more with research centres and universities to support investigations related to local water cultures (including collection of oral history and related visual displays, social water management, good practices of transboundary cooperation in the water sector etc.) in order to integrate existing displays with different and more rich cultural perspectives – including traditional knowledge and techniques, and taking into account also indigenous wisdom and worldviews (UNCCD, 2005; UNESCO, 2006).

Also riverine communities should be more involved through participatory approaches in order to contribute to the reconstruction of local 'water worlds', as well as to shape water museums' key messages related to SDGs, that is to keep the message alive and uphold the sense of place bonded to local historical hydro-scapes, which link past and present, natural and cultural, tangible and intangible water heritage. Today new interdisciplinary and holistic approaches are essential to frame a stronger strategy to promote and disseminate the urgent actions which are needed to preserve both natural and social water assets for future generations.

13.5 CONCLUSIONS

Hydro-scapes which once were flourishing and at the very centre of social life are today quite often abandoned, and thus made 'invisible' both to residents and to potential tourists. However, neglected hydraulic heritage has today an unparalleled chance of redemption if the use of digital media, applied research investigation and the involvement of local communities are properly combined. Research is necessary not only to digitalize hydro-scapes at local and regional level, but also to settle the right approach and strategy for regeneration of the forgotten heritage.

Digital media, and in particular geospatial technologies, are fundamental to facilitate and promote access towards this invisible heritage and to help to boost new eco-tourism and craft tourism activities. Specific techniques to facilitate participatory approaches are useful to involve residents and local associations and make them aware of the potential role they can play in this process. As a final output, it is possible to assist to the regeneration of neglected rivers and waters.

Also youth involvement in water awareness educational activities linked to the inherited hydro-scapes can help to stimulate an increased comprehension of the different water uses and their real sustainability in the years to come, therefore accomplishing the targets of SDGs. The role of young people, residents and local associations is crucial to restate the value of the inherited 'liquid heritage' and to influence current dynamics of water management and territorial planning at regional level – those dynamics where environmental, socio-cultural and hydraulic aspects interact (Cosgrove & Petts, 1990; Visentin, 2014).

As the case of Venice's inland waterways demonstrates, applied research investigation not only helps to understand the evolution of local water flows' management dynamics, but also the possible rediscovery of the local 'sense of place' by riverine communities which, in recent decades, have been deeply impacted by the predominant mind-set aimed at achieving only a standardized cultural and economic development: a development which does not consider and give value to the watery *genius loci*, and therefore exclude sustainability by definition – a kind of development that unfortunately produced, in the end, the same final decay of Venice's inland hydro-scapes, as opposed to other European regions.

Today, Venetian waterways' heritage is highly valuable not only for its century-old tangible and intangible cultural assets, including monumental architecture, knowledge and craftsmanship related to inland nautical culture, but also for its ecological value and high potential to increase biodiversity. Growing social demands for green spaces and attractive waterscapes should be considered indeed as an undeniable expression of a new 'hydraulic humanism' (Vallerani, 2017). In this perspective, by combining both cultural and natural components, the water heritage could be just functional to promote everywhere a 'new water culture' through alternative tourist experiences and water awareness educational approaches, where the recovery of hydro-scapes goes hand-in-hand with UNESCO's targets of SDGs.

The revaluation strategy developed by the Water Museum of Venice to make more easily accessible a forgotten hydraulic heritage through digital media and the involvement of local communities, with the methodological premises and project steps discussed in this chapter, may serve as a model to stimulate additional processes of re-valuation of the 'invisible' hydraulic heritage not only in Europe, but anywhere. In this process, water museums can play a leading role in achieving SDGs. This is the challenge acknowledged by the Resolution XXIII-5 of UNESCO-IHP, which established in 2018 the Global Network of Water Museums to support water sustainability education and new water awareness efforts.

Nowadays different museums are engaged all over the globe in implementing water awareness educational activities which emphasize new possible links among past and present water civilizations, that is, among the inherited water culture and the more recently integrated water management models aimed at strengthening sustainability. This is a perspective where even the minor or secondary hydro-scapes, whether overlooked or made invisible in recent decades, can have a new role for local development and be redeemed, above all, by increasing awareness about the inherited water values which have been shaped over many centuries. In this frame, water museums' different multidisciplinary approaches definitely need more coordination at global level, in order to strengthen the new holistic perspectives which are necessary to stimulate and establish a 'new water culture'.

REFERENCES

Arnold D. and Geser G. (2008). Research Agenda for the applications of ICT to cultural heritage. Archaeolingua, Budapest.

Buhalis D. (2003). ETourism: Information Technology for Strategic Tourism Management. Pearson, London.

Chang H. (2015). Which one helps tourists most? Perspectives of international tourists using different navigation aids. *Tourism Geographies*, **17** (3), 350–369.

Ciriacono S. (2006). Building on Water. Venice, Holland, and the Construction of the European Landscape in Early Modern Times. Berghahn Books, Oxford.

Civiltà dell'Acqua International Centre. (2014). Water Museum of Venice [online] available at: http://www.watermuseumofvenice.com/ [accessed 13 January 2019].

Cosgrove D. (1990). Platonism and practicality: hydrology, engineering and landscape in sixteenth-century Venice. In: Water, Engineering and Landscape, D. Cosgrove and G. Petts (eds.), Belhaven Press, London and New York, 35–53.

Cosgrove D. (1993). The Palladian Landscape: Geographical Change and its Cultural Representations in Sixteenth Century Italy. Pennsylvania State University Press, Philadelphia.

Cosgrove D. and Petts G. (eds.) 1990. Water, Engineering and Landscape. Belhaven Press, London and New York.

Cucchini E. (1931). La navigazione interna dell'alta Italia [Inland navigation in Northern Italy]. Provveditorato Generale dello Stato, Roma.

D'Alpaos L. (2010). L'evoluzione morfologica della laguna di Venezia attraverso la lettura di alcune mappe storiche e delle sue carte idrografiche [The morphological evolution of the Venice Lagoon through historical maps and hydrographic cartography]. Comune di Venezia, Venice, Italy.

Dickinson J. E., Ghali K., Cherrett T., Speed C., Davies N. and Norgate S. (2014). Tourism and the smartphone app: capabilities, emerging practice and scope in the travel domain. *Current Issues in Tourism*, **17**(1), 84–101.

Eulisse E. (2014). Glimpses of a new water civilisation. In: Water Scarcity, Security and Democracy: A Mediterranean Mosaic, F. De Chatel, G. Holst-Warhaft and T. Steenhuis (eds.), Global Water Partnership Mediterranean, Cornell University & the Atkinsons Centre for a Sustainable Future, Athens & Ithaca NY, 20–33.

Eulisse E. and Visentin F. (2018). Digital applications and river heritage: the inherited landscape of Venice's historic waterways. In: Waterways and the Cultural Landscape, F. Vallerani and F. Visentin, Routledge, London.

European Commission. (2014). Towards an integrated approach to cultural heritage for Europe. [online] available at: http://ec.europa.eu/assets/eac/culture/library/publications/2014-heritage-communication_en.pdf [accessed 10 January 2019].

Franzin R. (2005). Il respiro delle acque [The breath of waters]. Nuova Dimensione, Portogruaro, Italy.

Gilchrist P. (2017). 'Where do heritage trails go to die?' Stepping out at the British seaside. In: Tourism and Heritage in Britain and Ireland: Conservation, Nation, Identity, G. Hoope (ed.), Palgrave, Basingstoke, 195–211.

Global Network of Water Museums (2018). Available at https://www.watermuseums.net/ [accessed 18 April 2021].

Hassan F. (2010). Water. History for our times, IHP Essays on Water History. UNESCO International Hydrological Programme, UNESCO Publishing, Paris.

Haus G. (2016). Cultural heritage and ICT: state of the art and perspectives. *Scientific Journal on Digital Cultures*, **1** (1), 9–20.

Kalay Y. E., Kvan T. and Affleck J. (eds.) (2008). New Heritage: New Media and Cultural Heritage. Routledge, London.

Lane F. (1973). Venice, a Maritime Republic. The Johns Hopkins Press, Baltimore.

Neuhofer B., Buhalis D. and Ladkin A. (2012). Conceptualizing technology enhanced destination experiences. *Journal of Destination Marketing & Management*, **1**, 36–46.

Norwich J. J. (1981). Venice: the Greatness and the Fall. Allen Lane, London.

Pease W., Rowe M. and Cooper M. (eds.) (2007). Information and Communication Technologies in Support of the Tourism Industry. Idea Group Publishing, London.

Pergolis R. and Pizzarello U. (1981). The Boats of Venice. L'Altra Riva, Venezia, Italy.

Prideaux B. and Cooper M. (eds.) (2009). River Tourism. CABI, Cambridge, MA.

Schama S. (1995). Landscape and Memory. Alfred Knopf, New York.

Shipley R. (1999). Making a silk purse from a sow's ear: lessons from the failure to develop the Welland Canal as a tourism resource. *Environments*, **27**(2), 25–44.

L. Smith and N. Akagawa (eds.) (2008). Intangible Heritage. Routledge, London.

Swyngedouw E. (2015). Liquid Power. Contested Hydro-Modernities in Twentieth-Century Spain. MIT Press, London and Cambridge, MA.

Tang L. and Jang S. (2010). The evolution from transportation to tourism: the case of the New York canal system. *Tourism Geographies*, **12**(3), 435–459.

Tiepolo M. L. and Rossi F. (eds.) (2008). Il governo delle acque [The management of waters]. Istituto Veneto di Scienze, Lettere ed Arti, Venezia.

UNCCD. (2005). Promotion of Traditional Knowledge. UNCCD, Bonn.

UNESCO. (2006). Water and indigenous peoples. In: Knowledges of Nature 2, R. Boelens, M. Chiba, D. Nakashima (eds.), UNESCO, Paris.

Vallerani F. (2016). Modernism and water engineering as development strategies: the case of Venetian inland waterways in the New Italian State (1866–1966). *Revista Movimentos Sociais e Dinamicas Espaciais*, **5**(01), 184–204.

Vallerani F. (2017). Italian minor rivers in a bio-regionalist vision: the case of the low plain between the Euganean Hills and the Venetian Lagoon. In: Cultural Heritage. Scenarios 2015–2017, S. Pinton and L. Zagato (eds.), Creative Commons, 733–744.

Vallerani F. and Visentin F. (2018). Waterways and the cultural landscape. Routledge, London.

Visentin F., Vallerani F., Ribas A. and Pavon D. (2014). *Waterscapes and engineered hydrography in Veneto, Catalonia and South-West England: a comparative study*. In: Book of the Congress on Industrial and Agricultural Canals, J. Porta (ed.), Fundació Universitat de Lleida, Lleida, 209–214.

Wang D., Park S. and Fesenmaier D. R. (2012). The role of smartphones in mediating the touristic experience. *Journal of Travel Research* **51**, 371–387.

Willems W. and Schaink H. (eds.) (2015). Water & Heritage. Material, Conceptual and Spiritual Connections. Sidestone Press, Leiden.

Xiang Z. and Tussyadiah I. (eds.) (2014). Information and Communication Technologies in Tourism 2014. Springer, Berlin.

Zorzi A. (2011). Venice 697–1797. A City, a Republic, an Empire. Peter Mayer Publisher, Woodstock & New York.

Chapter 14

A museum that travels: Yaku Viajero and citizen education on Guayllabamba social basin in Quito, Ecuador (South America)

José Ricardo Martínez, Cristina Reyes and Emilia Vallejo

ABSTRACT

Yaku Viajero is a travelling exhibition created by the Yaku Water Museum in Quito, Ecuador that aims to raise awareness among citizens about the importance of water conservation for the city and the world. It deals with five topics: water history in Quito, water quality, the right to water according to the constitution of Ecuador, Water Footprint and the social and hydrographic basin of the Guayllabamba, which provides water to the city of Quito, capital of Ecuador. This chapter describes in depth how the concept of the social basin is applied through educational and museographic resources and how it aims to promote the development of empathy as a fundamental basis of citizenship education.

14.1 INTRODUCTION

In December 2005, Yaku, the Water Museum, was inaugurated, as a cultural space conceived with the intention of educating citizens about the responsible use of water and presenting the theme from an artistic context. To date, its mission is to be a space that detonates more questions than answers; through reflection, participation and recreation, and the assumption that life depends on water. It bases its work on two thematic axes, water and nature, and water and society, addressing the water issue through non-formal education.

 The Cultural Dynamics in Water Management from Ancient History to the Present Age
Editor: Xiao Yun Zheng
doi: 10.2166/9781789062045_0241

Since 2008, the work of the museum has been legally supported by the Constitution of the Republic of Ecuador, which states, in the first section, Water and Food, article 12: 'The human right to water is fundamental and inalienable. Water constitutes a strategic national patrimony for public use, inalienable, imprescriptible, unattachable and essential for life, on this constitutional basis).' Therefore, it should be noted that we are a country that has articles in its constitution where the importance to the protection of water and natural resources is given.

Among all the issues that Yaku has worked on since it opened its doors to the public there is a particular issue that has been addressed on several occasions and with different methodologies: this is the Guayllabamba watershed and we are currently expanding this issue in conjunction with the Social Basin, involving the participating public as active actors and making their relationship with it visible (Fassi, García y Peralta, 2008).

The hydrographic basin of the Guayllabamba is located in the city of Quito, capital of Ecuador, and takes its name from the main river that crosses it. It is surrounded by eastern and western mountain ranges, and when the ice and snow on these mountains melts in the sun, they form a watershed (or drainage divide), where the water drains to form the different rivers that cross Quito, supplying water to the entire population. This makes up the physical or geographical basin; however, on the banks of the rivers, there are diverse populations that are dedicated to agriculture, livestock farming and other activities, related directly or indirectly to the water of these rivers. We will refer this interrelation of the people with the water as the social basin.

We decided to include the issue of Watersheds and Social Watersheds in the travelling exhibition 'Yaku Viajero' dealing with the idea 'water comes and water goes'. This is a theme that first invites us to place ourselves within our geographical context as citizens of Quito and then to know the snow-capped mountains which surround us and to value them, by raising awareness and reflecting on the care and conservation of these sources of liquid water, so vital to life.

The way we treat the water in the upper parts of the hydrological system, that is to say, the rivers, so that it is returned clean to the lower parts, that is to say, the oceans, is a subject that invites us to raise awareness that we are all part of the social basin and therefore it is up to everyone to take care of it.

14.2 THE CASE STUDY

14.2.1 Some theoretical concepts that are found in the development of contents of this topic

A *hydrographic basin* is the area of runoff of rainwater to a river or certain point. Its outline or perimeter is limited by the summits or edge of the mountains, and is called water parting, divortium aquarum or watershed. This watershed marks the boundary between one basin and neighboring basins. The water that falls on each side of the divide usually flows into different rivers. The term 'hydrographic' is related to the area that is mapped for the analysis of the components of the basin.

Therefore, a *basin* or *river basin* is defined as a specific segment of the Earth's surface, delimited from adjacent segments by more or less well-defined physical boundaries, also called dividing lines of the basin, and occupied in a period of time by plant and animal groups, including humans and interacting with the non-living components of the compartments, such as water, soil and air. Consequently, the concept of the ecosystem and its theory provides the theoretical framework for the adequate management of our natural resources and, therefore, of a basin or river basin.

From a more utilitarian perspective Parra 1988 (as cited in Desarollo de contenidos recurso educativo Cuencas Hidrográficas), we can consider a hydrographic basin as an integrated system to transform various environmental-geographical factors (topography) and capital investment into forest and agricultural products, wildlife, aesthetic and recreational satisfactions, and production of energy and water for the population, agriculture and industry. In this way, a basin is an ecosystem that serves society.

The basin as a unit has geographical, physical and biological characteristics that make it function as an ecosystem. This is why watersheds are the best geographical unit for regional planning, known as Watershed Management.

A river basin has a main river that empties into the sea, and it collects water from tributaries. The tributaries are secondary rivers that in turn form sub-basins. The relief of a basin consists of the main and secondary valleys, with the major and minor relief forms and the fluvial network; there are also mountains and their flanks; by the ravines or torrents, valleys and plateaus. The watershed has a concave shape where water drains into rivers and streams; the pattern of drainage in the basin closely resembles that of a leaf, since it shows that the side ribs resemble the tributaries and the center rib is like the main river.

By dividing geographical space by watersheds, several objectives are met:

- Find sources of problems, such as soil erosion and pollution, more easily and prevent them or minimize damage.
- Identify specific problems, especially those that connected with the rivers, the material or chemicals that they transport, their flow or the amount of water.
- Identify cumulative effects due to the interaction of inappropriate land use processes or practices that become significant for biological populations and water quality.
- Coordinate actions with populations or actors upstream and downstream.
- Identify and manage some biological processes, land use, agriculture that are not governed by geographical administrative spaces.
- Generate computer models for climatological or fluid information.
- Managing or re-establishing ecological flows: flows that are capable of maintaining the functioning, composition and structure of the fluvial ecosystem that this channel contains in natural conditions.

14.2.2 The Inter-Andean Alley, Ecuador

The Inter-Andean Alley is situated between the two internal slopes of the mountain ranges. It has a width of less than 40 km and an altitude between 1600 and 3000 meters. The alley is a succession of basins (holes) separated by transverse branches called 'knots' with elevations between 3000 and 3400 meters YAKU, Parque Museo del Agua (2010).

Each basin is a hydrographic system connected to the Pacific Ocean or to the Amazon River. From the north to the south, they are the Carchi, Mira, Guayllabamba, Chimbo, Chanchán, Cañar, Jubones, Puyango, Catamayo, and Macará rivers connect to the Pacific, and the Patate, Chambo, Paute, and Zamora rivers flow to the Amazon.

14.2.3 Hydrographic basin of the Guayllabamba

The upper basin of the Guayllabamba River is located in the Inter-Andean alley in the province of Pichincha, Ecuador. Its administration is made up of five cantons: Quito, Mejía, Rumiñahui, Pedro Moncayo and Cayambe. The altitude of this basin ranges from 5,893 meters above sea level (peak of the Cotopaxi volcano) to 1,000 meters above sea level (level of closure of the basin). This basin is one of the most densely populated areas of Ecuador, its total population (urban and rural) is approximately 2.5 million inhabitants.

The main economic activities that take place in the upper basin of Guayllabamba are agriculture, livestock farming, agro-industry, industry, tourism and generation of hydroelectricity. In the eastern basins, where the basin is provided with a large percentage of water for human consumption, the main economic activities, in order of importance, are livestock farming, agriculture, fish farming and ecotourism.

The main tributaries of the Guayllabamba basin are the rivers: Pisque, Pita, Monjas, San Pedro, Machángara. The Guayllabamba River empties into the Pacific, changing its name to Esmeraldas (Figure 14.1).

14.2.4 Management of the Guayllabamba basin

The term 'river basin management' admits multiple definitions, depending on the conceptual perspective of who uses it. In its origins, watershed management actions were aimed at regulating the water regime of the basin to ensure the availability of water resources in the medium and long term.

The approach now needs to be radically modified. Currently, it is not only a matter of implementing measures to sustain or increase the water yield of the basin, but also to ensure the existence and productivity of soil, vegetation and fauna resources. In other words, the problem is to achieve the rational management of all the resources of the basin (Revista Terra Incógnita, 2016).

Figure 14.1 Diagram about the hydrographical basin of Guayllamba.

14.2.5 Social basin

A social basin is a space delimited by the sources of the watercourses and the high areas that protect and nourish them and extends to where the waters reach naturally and to where the water is conducted by the threads built by the societies. It can be said that it is a complex composition of the geographical basin and its zones of influence determined by water users. A social basin usually implies the integration of several geographical basins intertwined by the social fabric that human beings construct.

The concept of the social basin is based above all on the intrinsic and complex relationship between water sources and the human beings that use and manage it, especially in the Andean area. The water strands within the basin connect thousands of users inside and outside the basin.

The social basin concept is something new; it is being developed especially in the Andean area, with sociological and anthropological studies seeking more visible participation of the communities and populations that inhabit the basin, with adequate, responsible and above all equitable water management. With the development of this concept, efforts are underway to generate policies relevant to the management and distribution of water, in addition to promoting the environmental care of the different areas of the basin to safeguard water sources (Figure 14.2).

14.2.6 Pollution in the Guayllabamba River basin

The following compilation of information on pollution of the Guayllabamba watershed is based on the work of Bonifaz (2018), pp. 11–12, 38–42. We have utilized this study because it uses reliable sources and, due to its academic origin, it is impartial showing strong data that will later allow us to justify the importance of the educational work of the Yaku Water Museum within the social basin.

The elimination of municipal wastewater in the environment without adequate treatment is considered as the main source of water pollution. As is well known, contamination causes decrease of dissolved oxygen in the water, causing the flora and fauna to die of asphyxiation.

The upper basin of the Guayllabamba River has several inhabited areas that have historically impacted the state of the river; this can be evidenced in the alteration of its biodiversity. Urban expansion and population growth is a latent threat because it brings with it activities such as agriculture with the use of fertilizers that can reach water and present a risk to the balance of the ecosystem; livestock and intensive and prolonged grazing, which as well as eroding the soil, can cause contamination with pathogens present in manure. It has also been shown that activities related to auto mechanics, trout farming and mining affect the river systems of the area. Carvajal mentions that several studies show that areas of páramo that have fewer human interventions have healthy ecological parameters and better habitats in rivers (Bonifaz, 2018, p. 39).

As as quoted in Bonifaz (2018, pp. 40–41), EPMAPS (Empresa Pública Metropolitana de Agua Potable y Saneamiento), FLACSO-Ecuador (Facultad Latinoamericana de Ciencias Sociales) and SENPLADES (Secretaría Técnica

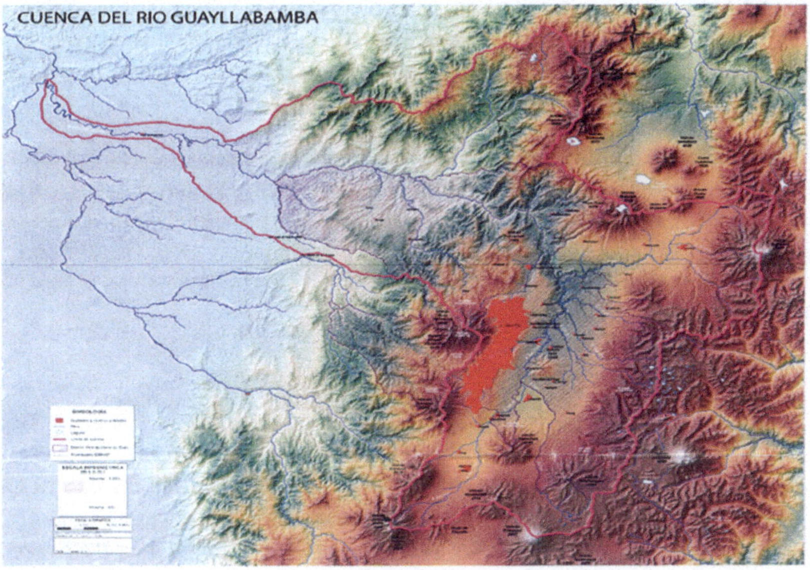

LIMITES

La cuenca del Río Guayllabamba se encuentra ubicada entre los nudos de MOJANDA al norte, que sirve de límite con la hoya del MIRA; y al sur con el nudo de TIOPULLO que sirve de límite con la hoya del PATATE. Al oriente sus límites son las cotas altas de la cordillera oriental (divorcios de agua), de la misma manera en el lado occidental, abriéndose paso el río Guayllabamba hacia la costa y terminando en su confluencia con el Río Blanco, formando con este el Río Esmeraldas.

Figure 14.2 Guayllabamba basin in Ecuador.

Limits

The Guayllabamba river basin is located between the Mojada knot to the north, which serves as the border with the Mira basin; and to the south with the Tiopullo knot that serves as a boundary with the Patate basin. To the east, its limits are the high elevations of the eastern mountain range, in the same way on the western side, the Guayllabamba river goes to the coast and ending at its confluence with the Blanco river, forming with it the Esmeraldas river with it.

Planifica Ecuador – Ecuador) collected the following information and data regarding pollution of the Guayllabamba river basin:

- In the year 2008, EPMAPS characterized wastewater discharges in several sectors of the metropolitan district of Quito. The average levels of the indicators were: Fecal coliforms, 8.36×10^7 cfu/l; biological oxygen demand, 302.4 mg/l) and chemical biological oxygen demand, 634.8 mg/l.
- Studies indicate that in 2011 the city of Quito consumed quantities close to 5.82 m^3/s in drinking water and generated approximately 4.8 m^3/s of wastewater. These waters are discharged mainly in the Machángara river with a flow of 0.69 m^3/s, and the Nuns river with flow at 0.64 m^3/s.

- EPMAPS conducted studies on water quality, which established in years prior to 2018 that the San Pedro, Machángara, Monjas and Guayllabamba rivers are not suitable for any use, such as the supply of drinking water, preservation of aquatic and wildlife, irrigation or recreational use.

One of the main challenges taken on by EPMAPS has been to decisively face the problem of contamination of the rivers that cross the Metropolitan District of Quito. In this connection at the beginning of 2011, the organization could count on a feasibility study of the plan for the decontamination of the rivers of Quito with a focus on the hydrographic basin.

The strategy aims to decontaminate the channels in:

(1) The Quebradas del Sur, Quito.
(2) Oriental Rural Parish of the valleys of Chillos and Tumbaco.
(3) City of Quito and annexed parishes.

EPMAPS (as cited in Bonifaz, 2018, p. 41) stated that until 2013 the city of Quito had more than 97% of potable water coverage, 91% of sewerage and 0% of wastewater treatment, which means that the basin of Guayllabamba needed to have decontamination programs. It has been proposed that the scheduled intervention for wastewater treatment should be implemented at the same time as the decontamination programs.

In 2020, Quito's first wastewater treatment plant began operating in the sector of Quitumbe and Beaterio with flows of 110 and 350 l/s, to serve a population of approximately 320,000 inhabitants. It is located to the south of the city, where the Machángara river rises. In addition, the construction of the Vindobona treatment plant is underway and is expected to be ready by 2025 and will treat 7.5 m^3/s from emissaries and interceptors that cover the entire flow of the city's effluents and channel it for approximately 34 kilometers; another solution is the 3-kilometer tunnel from the Nuns River in San Antonio de Pichincha to Vindobona, as well as a 20 MW hydroelectric plant that will take advantage of the existing difference of more than 350 meters between the plateau where the treatment plant will be located and the course of the Guayllabamaba River. These designs not only include dealing with sewage, but also the contaminated sludge generated as a result of the cleaning of the waters in order to return water of better quality to the Guayllabamba River (Vidal, 2014, pp. 365).

The information collected allows us to reflect on the misuse we make of water resources as a society and the importance of taking care of them, since all the inhabitants of the Metropolitan District of Quito have an important role in the care of water. Understanding the problem, we can face this with activities in which we feel part of both the problem and the solution (Moreano, 2015).

The decontamination plan recommends avoiding the disposal of solid waste to the water courses, that companies must have strict control of industrial discharges and appropriate management in the disposal of oils and used batteries, and above

Figure 14.3 Guayllabamba Valley (Machangara River)

all to promote effective environmental education that ensure citizen commitment to taking care of the streams and rivers, until they stop being considered as sewers or dumps so we can restore their landscape, their waters and their ecosystems (Figure 14.3).

14.2.7 From theory to educational resources to citizen education about the importance of the social basin

Yaku Viajero is a project that arises from the need of people who for various reasons cannot go to the Museum and request that we visit their communities with educational exhibitions, for which we designed a space that transcends its users so that we address issues of social interest that generate changes. One of its objectives is to offer visitors activities of dialogue participation and reflection for the construction of concepts and assessment of the watershed and social basin of the Guayllabamba River. The project uses interactive museographic resources, such as representations of mountains, rivers, which include illustrations of their flora and fauna, accompanied by texts and detonating questions that contribute to the construction of these concepts in a playful and participatory manner.

In the same way, we consider that knowledge of the hydrographic and social basin of Guayllabamba is very important since in this subject it is possible to observe closely which are the main sources of water supply and the relation of these with the people or communities that the surround them. We reach people dealing with these issues through games, educational museographic resources and above all using mediation, causing affective links to form between our visitors from different sectors and the travelling exhibition.

Below we describe two museographic resources, their interaction with the public and the results we have gathered from this experience.

14.2.8 Description of the resource/experience

The resource is a space in which the visitor will feel inside and part of the Guayllabamba river basin. The space has ten important images elevations of the basin, among them are snow-capped mountains such as El Pichincha, El Atacazo, El Cotopaxi where the rivers that flow through the city of Quito have their headwaters, for example, the Machangara, San Pedro and Monajas Rivers. These rivers become contaminated when they flow through the city. The mediator walks around the space with a group of approximately 15 people showing the images in the western cordillera and in the eastern ridge of the basin; this is intended to enable people to locate themselves geographically within the space and recognize the most representative flora and fauna of these elevations.

In addition, the mediator explains the source of each river rises because some rivers are still clean and others polluted; the mediator then explains how they join and form a single main river called Guayllabamba that crosses the basin and then joins the Esmeraldas River and finally ends in the Pacific Ocean (Junkin, 1985).

In other words, the mediator uses all the museographic resources to generate significant visitor experience with respect to the concepts and valuation of the river basin, its importance for the conservation and non-contamination of water.

14.2.9 Educational and museography experience

The resource demonstrates the water system of the Guayllabamba basin and how the pollution of each watershed affects the main river. Through the use of interactive forms, it provides knowledge of the species of plants and animals that inhabit the basin and their relationship with water sources (YAKU, 2017).

In Figure 14.5, 14.6 we show the museographic resources that we use to mediate the issue of the hydrographic basin.

Through a game of balance with several people, we talk about the social basin, the rules are to move a ball representing water through a labyrinth that has illustrations of the people that make up the social basin.

Failure to complete the task would mean losing the game because the players misused the water and as a result it could not reach other people or their destination (this is a real fact in various parts of the Guayllabamba basin, Ecuador and the world). When a group of visitors who are playing manages to place the ball in the center of the game, it shows that they did an excellent job and it is the right moment to reflect that the teamwork ensures that the water is returned clean to nature and that we are all responsible for taking care of the water and the watershed to which we belong.

Figure 14.4 Quito roads and the Machangara river. (With permission of J. R. Martinez).

In Figure 14.8, we show the museographic resources that we use to mediate the issue of the social basin (Figure 14.5–14.8).

14.2.10 The social basin and its relationship with the development of empathy and critical thinking about water as a common responsibility

When we created Yaku Viajero, we asked ourselves how to make people of all ages understand that clean and fresh water arrives in most of the city of Quito because of the work and support of several people and organizations. We also considered that this new knowledge will be in the minds of our visitors and will go beyond the idea of 'putting yourself in the shoes' of those people and institutions that contribute with work and effort in the day to day management and care of the basin of Guayllabamba.

It is for that reason, that with the resources created and already described, what we seek is to go beyond formal knowledge and memory of the information presented and to promote a real cognitive and emotional connection with the people behind the water supply to a city. The definition of empathy fitted pretty well with what we were looking for: according to the Royal Spanish Language Academy empathy means 'the ability to perceive, share and understand (in a common context) what another being can feel.' Another description of empathy given by Immordino Yang & Damasio is: 'affective participation of one person when another person is affected' (2017, p. 501). With the work of our educational mediators and through the dialogue and gaming. Yaku Viajero tries to kindle in our visitors emotion and affection so that through these feelings the learning becomes effective, as

Figure 14.5 Museographic design of educational resources about the basin (Yaku, 2017).

described by Mary Helen Immordino Yang (2007) in her investigations about the intimate relationship between those we feel and what we learn. Neuroscience research highlights connections between emotion, social functioning and decision making that change the dynamic of the role of affect in education. According to Immordino-Yang (2007), the neurobiological evidence suggests that the aspects of cognition that we encourage most heavily in schools, namely learning, attention, memory, decision making, and social functioning, are all profoundly affected by and subsumed within the processes of emotion called **emotional thought.**

Figure 14.6 Educational design of resources about the animals and plants from the basin (Yaku, 2017).

New research from neuroscience applied to education and culture tell us that emotion is the ideal means for cognition become embedded in the human mind and new ideas become transcendental and significant. In the words of Immordino-Yang (2007):

> Emotion, then, is a basic form of decision making, a repertoire of know-how and actions that allows people to respond appropriately in different situations. The more advanced cognition becomes, the more high-level reasoning supports the customization of these responses, both in thought and in action … The more people develop and educate themselves, the more they refine their behavioral and cognitive options. In fact, one could argue that the main purpose of education is to cultivate children's building of repertoires of cognitive and behavioral strategies and options, helping them to recognize the complexity of situations and to respond in increasingly flexible, sophisticated, and creative ways. (p. 7)

This is the educational base of Yaku Viajero and the way in which museology and museography has solved the idea of raising awareness about the importance

Figure 14.7 The river and mountains part of Guayllabamba basin in Yaku Viajero (Yaku, 2018).

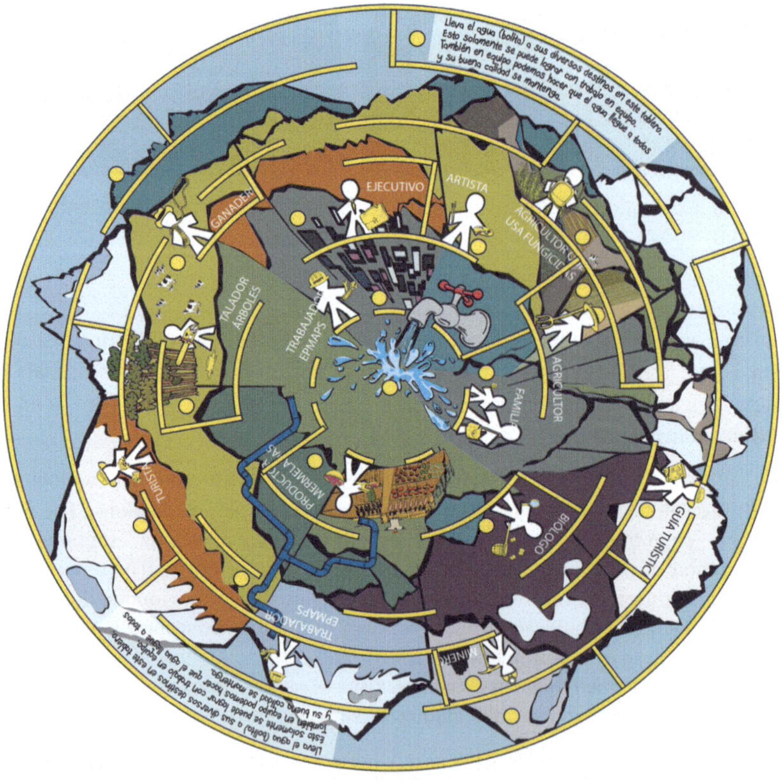

Figure 14.8 The social basin at Yaku Viajero (Yaku, 2017).

Figure 14.9 The educational resource showing the social basin with a group of students (Yaku, 2018). (With permission of J. R. Martinez).

of teamwork through the empathy of the social basin in Guayllabamba. We expect our visitors (children and adults) to understand the concept of social basin, but also, that they feel connected emotionally with gratitude for the effort and work of each collaborator or institution present in the basin until the water reaches each of them, and that through empathy, the culture of water for care and protection is dynamic, transforming and allows small and large actions in our city and the world.

14.2.11 Outcome

The travelling exhibition Yaku Viajero has visited low-income schools and rural areas of our country. In May and June 2017 the exhibition benefited approximately 1,313 people, both adults and children. The exhibition tour fulfilled the objective of providing education that deals with issues related to the care and conservation of water, including the very important matter of the watershed and social watershed. This topic has generated much reflection, dialogue and participation by visitors, especially when they were working in a team for a common good (in this case the care and conservation of water environment). The visitors considered concepts such as we are all part of the watershed and social watershed and therefore responsible for returning clean water that comes from the mountains to nature, and to take care of the rivers do not throw chemicals in them because there are more people living downstream who need the life-giving liquid.

It should be noted that people who live in rural areas of the city of Quito, often for economic or physical reasons cannot access the museum facilities so do not have a chance to learn about the subjects they cover. It is for this reason that Yaku decided to create this travelling exhibition 'Yaku Viajero' on the theme 'water comes and water goes', using educational museographic resources that through mediation help to reflect, question and discuss issues related to the responsible consumption of water, quality of water we drink and main sources of supply.

REFERENCES

Bonifaz E. (2018). Evaluación de la Contaminación Fecal de Origen Humano y Animal en la Cuenca del Río Guayllabamba Mediante Indicadores Virales, pp. 11–12, 38–42.

Desarollo de contenidos recurso educativo Cuencas Hidrográficas. Yaku, Parque Museo del Agua (2010). (Unpublished document).

Empresa Pública Metropolitana de Agua Potable del DMQ. https://www.aguaquito.gob.ec/ (accessed May 6, 2019).

Fassi F., García D. and Peralta J. (2008). Edición: Interjuntas y Foro de Recursos Hídricos; Entre Kishkas y abogados, ¿Cómo tramitar derechos de agua? La experiencia de Interjuntas Chimborazo.

Guión Museológico educativo de la exposición Itinerante Agua viene agua va. YAKU Viajero (2017). (Unpublished document).

Inmordino-Yang H. and Damasio A. (2007). We feel therefore we learn: the relevance of affective and social neurocience to education. *Mind, Brain and Education*, **1**, 3–10.

Junkin E. (1985). 'Agua y Salud Humana', Organización Panamericana de la Salud, oficina Sanitaria Panamericana, oficio Regional de la Organización Mundial de la Salud. Perú.

Moreano M. (2015). El agua de Quito. Revista Terra Incógnita. Quito, Ecuador, pp. 12–20.

Revista Terra Incógnita (2016). Por las vías del tren, más de 100 años de historia agro diversidad las rutas del tren oficios y artesanías, p. 30–38.

YAKU, Parque Museo del Agua (2010). 'Programa Gestión Museológica Exposiciones Temporales H2Q, gente por el agua.

Index

IWA Publishing's authorised EU representative for General Product Safety Regulations is Diane D'Arras, 15 rue Duret, 75116 Paris, France, e-mail: safety@iwap.co.uk.

Printed and bound by CPI Group (UK) Ltd, Croydon, CR0 4YY
12/05/2026
02108877-0002